Annals of Mathematics Studies
Number 65

LECTURES ON BOUNDARY THEORY FOR MARKOV CHAINS

BY

KAI LAI CHUNG

WITH THE COOPERATION OF

Paul-André Meyer

PRINCETON UNIVERSITY PRESS
AND THE
UNIVERSITY OF TOKYO PRESS

PRINCETON, NEW JERSEY
1970

LC Card: 72-106391
ISBN: 0-691-08075-5
AMS 1968: 6065

Published in Japan exclusively by the
University of Tokyo Press;
in other parts of the world by
Princeton University Press

Printed in the United States of America

PREFACE

These lectures were given at the Université de Strasbourg between November 1967 and March 1968 during a sabbatical leave. I had prepared a set of notes on selected and organized portions from [2] and [3] together with pre-requisites from [1] in a considerably simplified version. Professor Paul-André Meyer attended these lectures and took notes but wrote them up in his own style. He made many improvements in substance as well as in form. It afforded me genuine pleasure to see the material remolded by such a fresh and capable hand, and these notes are now presented here with only minor changes. Even when Professor Meyer's sentences sound a bit more French than English, no editor's pencil has been allowed to spoil the original flavor of his writing. The Prologue has been added at the completion of the manuscript.

The boundary theory as viewed here is a study of Markov chains with certain infinities (discontinuities that are not jumps) in the sample functions. It has two connected but distinct components. In the first part a simple "boundary" is set up by a direct, intuitive method and the basic relevant sample behavior is studied. Here a more complete treatment would perhaps imbed the chain in a fairly standard process on a state space which is some compactification of the countable, discrete space of the chain. There are those who seem to hold that such a possibility supersedes further work in this direction, presumably in the same sense that the advent of Čech, for instance, renders obsolete questions of convergence and continuity that pestered Euler, Cauchy and Weierstrass. I find it, however, more interesting to search for new structures in a relatively tame chain than to cover up the traces by ponderous escalation. The two approaches should reach some kind of mutual understanding soon if greater progress is to be made in the field.

The second, more substantial part of these lectures deals with the analysis of the chain in relation to its minimal part and its exits and entrances to and from the boundary. Historically this arose from the problem posed by Feller (1957) to construct all Markov chain semigroups from a given initial derivative matrix. Namely, given $Q = (q_{ij})$ subject to certain hypotheses, to find $\Pi = (p_{ij})$ such that $p'_{ij}(0) = q_{ij}$ for all i and j. This analytical problem is treated here by studying the "movements" of the underlying process. However, only the *better* half of the so-called "complete construction theorem" is given (see p. 72), which exhibits the necessary form of any semigroup that is a solution of the problem. The other half, to the effect that any similarly constructed form does yield a solution, is not discussed here. The latter is a somewhat tedious though not uninteresting algebraic exercise on Laplace transforms and may be found together with some serious discussion in §§16 and 18 of [2]. This omission is mentioned here since for other authors on the subject such a result has been the *pièce de résistance*, while for me the process is the thing. For the same reason *pro forma* citations of the literature have been kept to a minimum, but some historical references and comments may be found in [1], [4] as well as in the Prologue, and a very short list of papers in the Bibliography.

During my sojourn in Strasbourg I enjoyed the hospitality of the Départe ment de Mathématique in a brand new building with its excellent facilities. I had the privilege of counting not only M. Meyer but also MM. Cartier, Dellacherie, Morando, and Michel Weil in my audience. It is a pleasure to take this occasion to thank them and other members of the Département for their courtesy and assistance. The Office of Scientific Research of the United States Air Force, through a research contract, supported in part my sabbatical leave and also the preparation of the manuscript. Arthur Pitteng helped me in the preparation of the preliminary sections of the notes; Catherine Doléans, Gaston Giroux and Charles W. Lamb read the manuscrip and corrected many slips. Miss Gail Lemmond did the expert typing.

To them all I owe sincere thanks. My wife and son spent three months in a little apartment over a patisserie and shared with me the delights and frustrations of a transient, boundary existence. This little volume will be a memento to those eventful days.

Kai Lai Chung

PROLOGUE

If a continuous parameter Markov chain assumes only a finite number of states, almost all its sample paths can be described as follows. Starting at a state i, it will remain there for a sojourn time distributed with the density $q_i e^{-q_i t} dt$, $0 \leqq q_i < \infty$. Unless $q_i = 0$ the path will eventually leave i and jump to another state j with probability q_{ij}/q_i so that $q_i = \Sigma_{j \neq i} q_{ij}$. It will remain in j for a sojourn time distributed with density $q_j e^{-q_j t} dt$ and independent of the previous sojourn time in i, after which it will jump to another state k with probability q_{jk}/q_j, and so on. This is continued inductively until either the path enters an absorbing state and remains there forever, or makes an infinite sequence of sojourns-and-jumps such that the sum of the successive sojourn times increases to infinity. Thus each path is just a step function in every finite time interval and there are no other discontinuities other than isolated jumps. Analytically, the transition semigroup P(t) satisfies a pair of differential equations:

$$P'(t) = QP(t), \qquad P'(t) = P(t)Q \tag{1}$$

with

$$Q = P'(0) \ . \tag{2}$$

These are called Kolmogorov's backward and forward equations. respectively. They can be derived from the semigroup relation

$$P(s + t) = P(s)P(t) \tag{3}$$

by differentiating with respect to s and then setting $s = 0$. It follows that $P(t) = e^{Qt}$. The legitimacy of these steps must be proved, of course. Note that the matrix Q has nonpositive diagonal elements and nonnegative elements elsewhere, such that each row sum is zero, and that the q_i above is $-q_{ii}$.

The story for a finite-state Markov chain can be told rapidly from here on. It is not an exciting one from the probabilistic point of view, since a step function acts like a sequence of numbers. One might as well use a discrete time scale and reduce the analytic work to classical theory for positive matrices attached to the names of Frobenius and Perron. Indeed in the older literature the main concern is the ergodic type of limit theorems for $t \to \infty$.

When the number of states is (countably) infinite, these matrix methods become inadequate. As usual (in mathematics as in politics), efforts were made to cope with the new by the old, provided stringent restrictions were imposed. Such a condition is for instance that the Q matrix be bounded elementwise, or equivalently that the convergence of $P(t)$ to the identity matrix as $t \to 0$ be uniform in all elements. It turns out that under either condition almost all sample paths will again be step functions in every finite interval, just as in the case described above. The Poisson process and various compounds thereof fall in this category. Their great usefulness cannot be doubted judging from the annual production of papers on the subject, but it evades the new issues of a process running in continuous time and discrete space.

What can happen in general is as follows. First of all, some or even all the q_i's may be infinite. Such a state is called "instantaneous" and if a process starts from there it will go "in and out of" the state an infinite number of times (indeed on a set of positive measure) in any short time. Barring this, we suppose that all states are "stable" so that each will have an exponentially distributed sojourn time. However, upon exit from any of them the path may not jump to any other state but instead hit an infinity of them in arbitrarily short time. This happens whenever we have $q_i < \Sigma_{j \neq i} q_{ij}$, the "deficiency" $1 - \Sigma_{j \neq i}(q_{ij}/q_i)$ being equal exactly to the probability of a "pseudo-jump to infinity". If this is again barred then the deficiency must be zero for all i and the corresponding Q-matrix will be called "conservative". In this case the sample paths can indeed be traced through successive jumps as in the case of a finite number of states or the Poisson process, but only for an initial segment of time (which may be arbitrarily short for some samples). For now the sum of the sojourn times may converge to a finite limit, say $\tau = \tau(\omega)$. This is hardly surprising since the sum of the mean sojourn times, being just $\Sigma_i (1/q_i)$, can certainly converge in which event the Borel-Cantelli lemma implies at once the asserted result, almost surely. At this time τ, infinitely many jumps have accumulated to the left and the path goes to $+\infty$ if the states are the positive integers. Thus there is a discontinuity of the second kind, vulgarly described as an "explosion".

This last phenomenon, still dubbed as "pathological" by some authors, was actually discovered in a slightly devious manner. It should be borne in mind that three decades ago stochastic process was scarcely a mathematical subject (despite Wiener, Lévy and Doob) and it was even more popular then to solve differential equations than to trace sample paths. The equations in (1) present a neat analytic problem if Q is regarded as known and $P(\cdot)$ the sought solution. It was expected that if a nice solution could be found it would turn out to be a transition probability function. Now Feller in 1940, using an iterative method tantamount to the successive tracing of the jumps, succeeded in constructing such a solution $\Phi(\cdot)$ for a much larger class of Markov processes. It has all the properties of a transition function except that $\Phi(t)\,1 < 1$ is possible (in fact if this is true for any $t > 0$ then it is true for all $t > 0$). To quote Feller himself: "This is a surprising result and requires a better understanding of the mechanism of the process." If the solution Φ is defective, an immediate question arises as to whether there exists a true one? Doob (1945) was the first to find such a solution by looking at the paths. Is it not a matter of continuing it beyond the τ above, and if so what can be simpler than to re-start the path immediately after explosion at some favored state i (or a mixture of states)? The Markovian character must be preserved but that should take care of itself if one just lets the path, after re-starting, follow its own destiny as before, including the possibility of hitting infinity again and then re-starting again at the same i. This is precisely what Doob did except that he complicated matters unnecessarily by a transfinite induction. In the language of this monograph, Doob took the boundary to be a single atom and could have used $\eta_j(t) = f_{ij}(t)$ for all j as the Φ-entrance law. This solution satisfies the backward but not the forward equation. Indeed Doob gave necessary and sufficient conditions for a true solution to satisfy either equation, in terms of sample behavior which in the backward case reduces to the Q matrix being conservative as metnioned above.

The validity of Doob's construction can be easily verified by straightforward computation, but the "ghostlike return from infinity" was thought of as a feat bordering on the treacherous. It seemed also to have caused some confusion between a most particular example and a general analysis of the phenomenon. The latter was formulated as a problem of "complete construction" of transition semigroups under certain hypotheses. Now as any school child who has constructed a regular hexagon knows, the way to a construction problem is to begin by supposing that the required object has been given and try to analyse its properties until its *raison d'être* is discovered; to put it together then becomes a simple matter of retracing the steps. This is the natural point of view and is the one taken in this monograph. But first a bit more of history.

Paul Lévy (1951) found it all too easy to describe a whole hierarchy of pathological behavior of sample paths. Perhaps the most striking one is where the process assumes all the positive rational numbers as stable states, with sojourn times adding up to infinity as they are gone through in their natural order, while it spends almost no time in the intervening irrational numbers. Such a path has no jumps at all but only discontinuities of the worse kind. The corresponding Q matrix has finite diagonal elements but zero elsewhere. Intuitively trivial as the example is (after Lévy displayed it like Columbus's egg) it is a tenuous as well as tedious matter to verify the Markovian character of the process. At about the same time Kolmogorov published two examples of transition function, one of which establishes the possibility of an infinite pseudo-jump as described earlier, and the other the existence of an instantaneous state. His construction is purely analytic and led D. G. Kendall and G. E. H. Reuter to a scrutiny by semigroup methods and the construction of other bizarre examples.

But Lévy's work also opened up the field for positive advance. Coupled with some martingaling due to Doob (1942), it became possible to describe the basic topological properties of the sample paths in general (see Theorem 1 of Chapter II, Section 2 for the case of stable states). A form of the strong Markov property emerged. Reversing the historical trend, sample analysis was effectively used to establish analytic facts such as the differentiability of transition functions. Some of these developments will appear in this monograph but a more complete account with references is given in [1].

Here we shall concentrate on the specific setting above: namely all states are stable and there are no infinite pseudo-jumps. The analytic problem is then to find all transition semigroups $P(\cdot)$ satisfying (2) with a given conservative Q. Such a solution will automatically satisfy the backward equation as mentioned earlier, and it may be interesting to single out those which satisfy also the forward equation. This is the approach of Feller (1957) who regards the equations in (1) as a pair of adjoints with an apparent lack of symmetry, and constructs solutions by means of boundary conditions in analogy with classical theories of such differential equations. He introduces an exit and an entrance boundary and assumes that they be both finite in his final result. The last condition may be reduced to an algebraic formulation as follows. For a given $\lambda > 0$ (hence for all $\lambda > 0$), each of the systems below has only a finite number of linearly independent nonnegative solutions:

$$\lambda x_i = \sum_j q_{ij} x_j, \qquad \sup_i |x_i| < \infty;$$

$$\lambda y_j = \sum_i y_i q_{ij}. \qquad \sum_j |y_j| < \infty.$$

Neveu (1961) extended Feller's ideas in a synthesis by lattice methods.

David Williams (1964), after precursory work by Reuter, showed that the assumption on the entrance boundary (equivalently on the second system above) may be omitted.

Our approach here, based on [2] and [3], is different from the previous studies in that the Markov chain as a stochastic process rather than its semigroup is the main concern, while solution of the analytic problem above is obtained as a by-product. The two are closely related of course, but the process takes precedence, as will be demonstrated in the pages below. To return then to our interrupted tracing of the sample path, the question for us is what it does after the explosion. Since a reasonable version of the process is supposed to be given, each path is defined for all time and does not stop at τ. If it has gone to infinity there, it must perforce come back from it in a flash, since it necessarily assumes finite values at all rational times by the very definition of the process with the integers as its state space. It can of course go off to infinity again, perhaps in another flash, and return again, and so forth. This is indeed the case at a sticky atom of the boundary as defined in this monograph. Now, there are essentially different ways in which a path can escape to and return from infinity, and these must be considered as distinct ideal states, or boundary points in the terminology coined by Feller. Such ideal structures are known in all parts of mathematics under various names as ideals, completion and compactification. The creation of the real numbers through the rational is the most celebrated illustration, and Lévy's example mentioned above illuminates the analogy. The total time spent at the boundary has only zero Lebesgue measure, yet a boundary state must behave in some ways like an ordinary one. For instance, if the process is at such a state, the past and future should again be conditionally independent, thus extending the known form of the strong Markov property to include stopping at the boundary. There should be transitions to and from the boundary and these should obey the usual laws of composition. Finally the whole process should be describable in terms of exits to and entrances from the boundary, together with the movements clear of the boundary which is controlled by the minimal semigroup. In this way the setting up of the boundary will serve to elucidate the passages of time and space that would have been obscure without it. It is the purpose of this monograph to develop such an account under the main hypothesis that the exit boundary is finite. Nothing much is known yet beyond this case.

References [2] and [3] contain several topics which are not included here, such as: extreme laws, dual chains, forward equations, the construction problem, different decompositions and their relations.

Among the latest results on the subject we mention [7], [14], [16], Dynkin extended Feller's work under the same conditions as in [17] and [2], but defined an entrance boundary. Pittenger generalized and strengthened the results relating to the main decomposition in Chapter 3 below. Walsh completed the definition of the exit boundary used here and added an appropriate entrance boundary by probabilistic methods. Further work is in progress in several doctoral dissertations.

BASIC NOTATIONS

1) T is the closed half line $[0, \infty[$.

   T^0 is the open real line $]0, \infty[$.

   I is a denumerable set, with the discrete topology, and the σ-field of all subsets of I.

   I' is the one point (Alexandrov) compactification of I. We denote by ∞ the point at infinity added to I.

2) A *matrix function* Π on I is a family $(\Pi(t))_{t \in T}$ of $I \times I$ matrices, and we shall denote by $p_{ij}(t)$ the (i,j) coefficient of $\Pi(t)$. Let us consider the following properties:

   Property i. $\Pi(t) \geqq 0$ for all t (i.e., $p_{ij}(t) \geqq 0$ for all t, i, j).

   Property ii. $\Pi(s + t) = \Pi(s)\Pi(t)$ for all s, t (i.e., we have

$$p_{ij}(s + t) = \sum_k p_{ik}(s)p_{kj}(t) \text{ for all } i, j, s, t).$$

   Property iii. $\Pi(t)1 = 1$ for all t (i.e., $\Sigma_j\, p_{ij}(t) = 1$ for all t and i)

   Property iv. $\Pi(0) = I = \lim_{t \to 0} \Pi(t)$ (i.e., $p_{ij}(0) = \delta_{ij} = \lim_{t \to 0} p_{ij}(t)$ for all i, j).

These four properties all have names: i) is "*positivity*" ; ii) is often called "*the semi-group property*" ; iv) is "*standardness*". If we replace iii) by $\Pi(t)1 \leqq 1$, then the matrices are "*substochastic*" instead of "stochastic".

We shall use the following terminology, which is inconsistent from the logical point of view, but is consistent with the principle of giving the simplest names to the objects which are most often used.

If Π satisfies all properties i-ii-iii-iv, then Π is a *standard transition matrix* (function): abbreviated STM.

If Π satisfies only i-ii-iv, then Π is a *general transition matrix* (function): abbreviated GTM.

If Π is a GTM, and $\Pi(t)1 \leqq 1$ for all t, then Π is a *substochastic transition matrix* (function): abbreviated SSTM.

We have omitted the word "standard" from the last two definitions and abbreviations, since nonstandard matrix functions will never occur.

3) Row vectors are used in matrix notation to denote *measures on* I, and column vectors to denote *functions* on I. If μ is a measure, f a function, we still denote by μ, f, the row and column vectors defined by $\mu_i = \mu(\{i\})$, $f_i = f(i)$. The integral of f with respect to μ (provided it exists) is written as $\langle\mu, f\rangle = \Sigma_i \mu_i f_i$. Matrices are interpreted as *kernels on* I. All vectors and matrices have *finite* coefficients, unless the contrary is explicitly stated.

Matrix notation is very concise, but often unreadable. For instance if F is a function (column vector) $\Pi(t)F$ is a function, but we have no place to write the variable. In the same way, if we want to consider the function $p_{\cdot j}(t)$, and substitute for $\cdot$ a random variable X, we shall have the very cumbersome notation $p_{X,j}(t)$. The only escape is inconsistency: we shall introduce besides the matrix notation the usual kernel notation: the function $\Pi(t)F$ will also be denoted by $\Pi(t, \cdot, F)$, the measure $G\Pi(t)$ (G is a row vector, or measure) will be denoted by $\Pi(t, G, \cdot)$, and the coefficient $p_{ij}(t)$ will be denoted sometimes by $p(t, i, j)$. We shall always indicate explicitly such changes of notation below.

4) *Positive* always means "non-negative" ($\geqq 0$). If we mean > 0, we say "strictly positive" (for numbers). In the same way, *increasing* maans non-decreasing.

CONTENTS

CHAPTER I

ELEMENTARY PROPERTIES OF TRANSITION MATRICES

Organization. This chapter consists of 2 sections. In the first one, entrance and exit laws are defined, and we prove two analytical lemmas which are true for all GTM. In Section 2, we prove some analytical facts (very simple) specific to STM's, and the basic results concerning differentiability of the transition matrix.

Section 1. Two Basic Analytical Lemmas

In this section, Π is a GTM. We shall need several times the obvious fact that $p_{ii}(t)$ never vanishes $(p_{ii}(t) \geqq (p_{ii}(t/n))^n)$.

DEFINITION 1. *An entrance law is a family* $(G(t))_{t>0}$ *of positive row vectors (measures) on* I *such that*

$$G(s)\Pi(t) = G(s+t) \qquad (s>0,\ t \geqq 0).$$

The entrance law is normalized if $\langle G(t), 1\rangle = 1$ *for all* $t>0$. *In the same way, exit laws are families* $(F(t))_{t>0}$ *of positive column vectors (functions) such that*

$$\Pi(t)F(s) = F(s+t) \qquad (s>0,\ t \geqq 0).$$

In non-matrix notation: Let $g_i(t)$ be the i-coefficient of $G(t)$. Then $g_i(t)$ is finite and $\geqq 0$ for all t, and $\Sigma_i\, g_i(s)\, p_{ij}(t) = g_j(s+t)$ for all j, $s>0$, $t \geqq 0$. Normalization means that $\Sigma_i\, g_i(s) = 1$ for all $s>0$.

Examples: Set $g_j(s) = p_{ij}(s)$ $(s>0$, i fixed), $f_i(s) = p_{ij}(s)$ $(s>0$, j fixed); this defines an entrance law $G(s)$, an exit law $F(s)$.

Define $\Pi^*(t)$ as the transpose matrix of $\Pi(t)$, i.e., $p^*_{ij}(t) = p_{ji}(t)$. Then if G is an entrance law with respect to Π, its transpose G^* [($G^*(t)$ is the column vector with the same coefficients as the row vector $G(t)$) is an exit law with respect to Π^*, and Π-exit laws in the same way become Π^*-entrance laws by transposition. Thus there is full symmetry between entrance and exit, as far as GTM's are considered—but unfortunately the transpose of a *stochastic* matrix function is just a GTM, and therefore this symmetry cannot go very deep.

This is the first basic lemma.

PROPOSITION 1. *Let* F *be a* Π-*entrance law, with coefficients* $f_i(t)$ $(t > 0)$. *Then*

a) *all coefficients* $f_i(\cdot)$ *are continuous functions in* T^0, *and have a finite limit* $f_i(0+)$ *at* 0.

b) *Let* $F(0+)$ *be the row vector with coefficients* $f_i(0+)$. *Then* $F(0+)\Pi(t) \leqq F(t)$ *for all* t. *If* $F(t) = F\,\Pi(t)$ *for some positive row vector* F, *then* $F(0+) = F$. *(A similar statement holds for exit laws.)*

Proof: a) Let us show first that $F(t+)$ exists for $t \geqq 0$ (we don't prove yet that $f_i(t+)$ is finite).

We have $f_j(s+t) \geqq f_j(s)\,p_{jj}(t)$. Therefore, if t is sufficiently small

$$f_j(s+t) \geqq (1-\varepsilon) f_j(s) \qquad \text{for } all \text{ s.} \tag{1}$$

Take $r \geqq 0$, and choose two sequences $r_n \downarrow r$, $r'_n \downarrow r$ (with values strictly $> r$) such that

$$f_j(r_n) \to \limsup_{\substack{s \to r \\ s > r}} f_j(s), \qquad f_j(r'_n) \to \liminf_{\substack{s \to r \\ s > r}} f_j(s)\,.$$

By extracting subsequences, we may assume that $r_n \leqq r'_n \leqq r_n + t$ for all n, and therefore from (1).

$$\liminf_{\substack{s \to r \\ s > r}} f_j(s) \geq (1-\varepsilon)\,\Big(\limsup_{\substack{s \to r \\ s > r}} f_j(s)\Big)\,.$$

Hence a) is proved. In the same way, we get

b) $F(t-)$ *exists for all* $t > 0$.

From (1) above, we get

$$F(t+) \geqq F(t) \geqq F(t-) \tag{2}$$

$$(t > 0; \text{ also } F(0+) \geq F \text{ if } F(t) = F\Pi(t) \text{ for all } t > 0),$$

and from Fatou's lemma:

$$F(s+)\Pi(t) \leq F(s+t+) \qquad \text{(all } s),$$

$$F(s-)\Pi(t) \leqq F(s+t-) \qquad \text{(all } s > 0). \tag{3}$$

c) Let D_j be the countable set of discontinuities of $f_j(\cdot)$. The set $D = \cup_j D_j$ still is countable. If $s + t \notin D$ we have

$$F(s+)\Pi(t) \leqq F(s+t+) = F(s+t) = F(s)\,\Pi(t) \leqq F(s+)\Pi(t) \;.$$

Therefore $s + t \notin D \Longrightarrow F(s+)\Pi(t) = F(s)\Pi(t)$. On the other hand, $F(s+) \geqq F(s)$, and therefore $f_i(s+) > f_i(s)$ is compatible with $\Sigma\, f_j(s+)\,p_{jk}(t) = \Sigma\, f_j(s)\,p_{jk}(t) = f_k(s+t) < \infty$ if and only if $p_{ik}(t) = 0$ for all k. But for any $s \geqq 0$ we may choose t such that $s+t \notin D$, and we have seen that $p_{ii}(t) > 0$. Therefore we must have $F(s+) = F(s)$ for all $s > 0$. The same proof shows that if $F(t) = F\,\Pi(t)$, then $F(0+) = F$.

d) The proof that $F(t-) = F(t)$ for $t > 0$ is quite similar, and isn't given here. The only statement left to prove therefore is the finiteness of $F(0+)$, which follows from the relation $F(0+)\Pi(t) \leqq F(t+) = F(t)$.

COROLLARY. *The functions* $p_{ij}(\cdot)$ *are continuous on* T.

Indeed, we can take $f_j(t) = p_{ij}(t)$ in proposition 1 (i fixed). We shall see a more precise result later on, for STM's (proposition 4).

We come to the second basic lemma (Chung, Neveu). We shall need the following theorem (quoted below as *Fubini's theorem* on derivation).

LEMMA. *Let* H_n *be a sequence of monotone increasing functions on* T^0 *such that* $H = \Sigma_n H_n$ *is finite. Then* $H'(x) = \Sigma_n H_n'(x)$ *for a.e.* x *(Lebesgue measure).*

We shall not prove this theorem. It is very simple to deduce it from the following well-known facts: a) if J is any increasing function, then J' exists a.e., and is the density of the absolutely continuous part of the measure dJ; b) the absolutely continuous part of dH is the sum of the absolutely continuous parts of the measures dH_n. For a detailed proof see for instance Saks, *Theory of the Integral*, Warszawa-Lwów, 1937. p. 117.

PROPOSITION 2. *Let* G *be a positive row vector function on* T^0 *such that*

$$G(s+t) - G(t) = G(s)\Pi(t) \qquad (s > 0,\ t > 0). \tag{1}$$

Then there exists an entrance law F *such that*

$$G(s+t) - G(t) = \int_t^{s+t} F(u)\,du \qquad (s \geqq 0,\ t > 0). \tag{2}$$

(This implies by proposition 1 that G *is continuously differentiable.)*

Proof: Let $G'(s)$ denote the right hand lower derivative of G at s. (1) implies that G increases with s, and therefore G' is positive. We have

$$\frac{1}{\varepsilon}(G(s+\varepsilon) - G(s))\Pi(t) = \frac{1}{\varepsilon} G(\varepsilon)\Pi(s)\Pi(t)$$

$$= \frac{1}{\varepsilon}(G(s+t+\varepsilon) - G(s+t))$$

and therefore, from Fatou's lemma

$$G'(s)\Pi(t) \leqq G'(s+t) \ . \tag{3}$$

This implies in particular $g'(s+t) \geqq g_j'(s)\,p_{jj}(t)$, and therefore the *finiteness* of $g_j'(s)$, since otherwise g_j' would be infinite at all points of $[s, \infty[$

a) Let us now write Fubini's theorem on derivation[1] (applied to the relations $g_i(\cdot + t) = g_i(t) + \Sigma\, g_j(\cdot)\, p_{ji}(t)$). We get: *For every* $t > 0$, *there exists a set* $Z_0(t)$ *of Lebesgue measure* $m(Z_0(t)) = 0$, *such that for all* $s \notin Z_0(t)$

$$G'(s+t) = G'(s)\Pi(t) \ .$$

We transform this statement, using Fubini's theorem (on inversion of integrations) getting:

(4) *For* $s > 0$, *set* $Z(s) = \{t : G'(s)\Pi(t) \neq G'(s+t)\}$, *and set* $Z = \{s : m(Z(s)) > 0\}$. *Then* $m(Z) = 0$.

b) We shall first prove that, if $s \notin Z$ (i.e., if $m(Z(s)) = 0$), $Z(s)$ is *empty.*

Take $t \in Z(s)$. Then (3) implies that, for all k,

(5) $g_k'(s+t) \geq \sum_i g_i'(s)\, p_{ik}(t)$ with strict inequality for some $k = j$.

Take $r > t$, $r = t + u$. Then

$$g_j'(s+r) \geqq \sum_k g_k'(s+t)\, p_{kj}(u) \underset{(*)}{\geqq} \sum_{i,k} g_i'(s)\, p_{ik}(t)\, p_{kj}(u)$$

$$= \sum_i g_i'(s)\, p_{ij}(t+u) \ .$$

Now (*) must be a strict inequality according to (5), since there is strict inequality for $k = j$, and $p_{jj}(u)$ doesn't vanish. Therefore $t \in Z(s)$ implies $r \in Z(s)$ for all $r > t$, and this contradicts the fact that $m(Z(s)) = 0$. The only escape is $Z(s) = \emptyset$.

---

[1] If the reader wants to avoid the well-known theorem that $G'(\cdot)$ is Lebesgue measurable, he can just define $G'(s) = \liminf_{n \to \infty} n[G(s+1/n) - G(s)]$.

c) Now let u and v be any numbers > 0. We may write $u = s + t$, with $s \notin Z$ (since Z has measure 0). Then we have from b)

$$G'(s + t) = G'(s)\Pi(t) \quad ,$$

(6)

$$G'(s + t + v) = G'(s)\Pi(t + v) = G'(s)\Pi(t)\Pi(v)$$

and therefore $G'(s + t + v) = G'(s + t)\Pi(v)$. Otherwise stated: $G'(u+v) = G'(u)\Pi(v)$. Since u and v are arbitrary, $G'(\cdot)$ is an entrance law, and therefore continuous (proposition 1). Proposition 2 follows at once.

The following result is an important application of proposition 2 to excessive measures.

PROPOSITION 3. *Let* H *be a "purely excessive" measure, i.e., a positive measure on* I *such that*

$$\text{(1)} \qquad H\Pi(t) \leqq H \qquad \textit{for all } t\ ;$$

$$\text{(2)} \qquad \lim_{t \to \infty} H\Pi(t) = 0 \quad .$$

Then there exists an entrance law $F(\cdot)$ *such that*

$$\text{(3)} \qquad H = \int_0^\infty F(t)\,dt \quad .$$

Moreover, F *is unique.*

Proof: Define $G(s) = H - H\Pi(s)$; according to proposition 2, there exists an entrance law F such that

$$H\Pi(t) - H\Pi(s + t) = G(s + t) - G(t) = \int_t^{s+t} F(u)\,du \quad .$$

Let t tend to 0, s to ∞: $H\Pi(t)$ tends to H (proposition 1, statement b)) and $H\Pi(s + t)$ tends to 0 according to (2). Therefore we get (3). Applying

$\Pi(s)$ to (3) we get

$$H\Pi(s) = \int_s^\infty F(u)du$$

and therefore $F(s) = -\frac{d}{ds} H\Pi(s)$, from which the uniqueness follows.

Section 2. Differentiability

In this section, Π is a STM. However, it is important to notice that all results extend trivially to the *substochastic* case, since any SSTM can be transformed into a STM by adding just one absorbing state to $\mathbf{I}$.

The following statements tell the basic facts about the analytical behavior of the transition matrix. These results are deep, and we shall prove only much weaker statements. For complete proofs, see Chung [1].

a) *Lévy's Theorem* (first complete proof by Austin). The function $p_{ij}(\cdot)$ either is identically 0 in T^0, or never vanishes in T^0.

b) *Ornstein's Theorem.* The function $p_{ij}(\cdot)$ is continuously differentiable in T^0 (but there are examples where it isn't twice differentiable).

Ornstein's theorem doesn't state anything about differentiability at 0. This was solved much earlier by Doob and Kolmogorov. Here is the result:

c) The limit $q_i = \lim_{t \to 0+} (1 - p_{ii}(t)/t)$ exists for all i, but it may be equal to $+\infty$.

The limit $q_{ij} = \lim_{t \to 0+} (p_{ij}(t)/t)$ exists and is finite for all i and $j \neq i$.

We have $\Sigma_{j \neq i}\, q_{ij} \leqq q_i$.

We shall prove part of these results below, but we shall use the notations of this statement, and set the following definition.

DEFINITION 2. *The state* i *is stable if* $q_i < \infty$, *instantaneous if* $q_i = \infty$ (we shall see that i is absorbing if and only if $q_i = 0$).

These general results being stated, we come to the results we shall actually need, and prove.

PROPOSITION 4. *For all* $t \geqq 0$, $h \geqq 0$, *we have* $|p_{ij}(t+h) - p_{ij}(t)| \leqq 1 - p_{ii}(h)$.

Proof:
$$p_{ij}(t+h) - p_{ij}(t) = \sum_{k \neq i} p_{ik}(h)\, p_{kj}(t) - p_{ij}(t) + p_{ii}(h)\, p_{ij}(t)$$
$$= \sum_{k \neq i} p_{ik}(h)\, p_{kj}(t) + [p_{ii}(h) - 1]\, p_{ij}(t) \quad .$$

To conclude, we just remark that these two terms have opposite signs, and both have an absolute value smaller than $1 - p_{ii}(h)$.

To prove proposition 5 we shall need a lemma on subadditive functions.

LEMMA. *Let* ϕ *be a finite valued, subadditive function on* T^0:

$$\phi(s+t) \leqq \phi(s) + \phi(t) \qquad (s > 0,\ t > 0)$$

such that $\lim_{t \to 0+} \phi(t) = 0$. *Then the limit* $\lim_{t \to 0+} (\phi(t)/t)$ *exists (but can be* $+\infty$*), and is equal to* $\sup_t (\phi(t)/t)$.

Proof: Set $c = \sup_t (\phi(t)/t)$; obviously $\limsup_{t \to 0+} (\phi(t)/t) \leqq c$. Take any $c' < c$, and choose s such that $(\phi(s)/s) > c'$. Take any $t > 0$ and write $s = nt + h$, with n an integer $\geqq 0$, and $0 \leqq h < t$. Subadditivity then gives

$$c' < \frac{\phi(s)}{s} \leqq \frac{n\phi(t) + \phi(h)}{s} = \frac{nt}{s}\,\frac{\phi(t)}{t} + \frac{\phi(h)}{s} \quad .$$

Now let t tend to 0: nt/s tends to 1, and this inequality implies

$$\liminf_{t \to 0+} \frac{\phi(t)}{t} \geqq c' \; .$$

QED

PROPOSITION 5. *The limit* $\lim_{t \to 0+} (1 - p_{ii}(t))/t = q_i \leqq +\infty$ *exists for all* i.

Proof: Since $p_{ii}(t) \to 1$, we may just as well replace $1 - p_{ii}(t)$ by $\phi(t) = -\log p_{ii}(t)$. Now $p_{ii}(s+t) \geq p_{ii}(s)\, p_{ii}(t)$, and so ϕ is subadditive (and obviously finite valued). The result follows immediately from the lemma.

REMARK. The lemma implies that $\phi(t) = 0$ for all t if $q_i = 0$. Therefore $q_i = 0$ if and only if i is absorbing. Remark also that proposition 5 is true for GMT's.

The following result gives, for a *stable* state i, both Ornstein's statement and Doob-Kolmogorov's (but the results for non-stable states lie much deeper).

PROPOSITION 6. *Let* i *be a stable state. Then*

1) $p'_{ij}(t)$ *exists, is finite and continuous for all* $t \geqq 0$ *and* $j \in I$ *(this implies in particular the existence and finiteness of* q_{ij}*).*

2) $p'_{ij}(s+t) = \sum_k p'_{ik}(s)\, p_{kj}(t) \quad (j \in I,\ s > 0,\ t \geqq 0)$[1].

3) $\sum_j p'_{ij}(s) = 0 \quad (s > 0)$[1]; $\sum_{j \neq i} q_{ij} \leqq q_i$.

4) $\sum_j |p'_{ij}(t)| \leqq 2q_i$.

Proof: We know from the lemma on subadditive functions that

$$q_i = \sup_r \frac{-\log p_{ii}(r)}{r} .$$

Therefore we have $p_{ii}(r) \geqq e^{-q_i r}$ for every r. Take $0 \leqq s < t$; we apply this with $r = t - s$ and get

$$p_{ik}(t) - p_{ik}(s) \geqq p_{ii}(t-s)\, p_{ik}(s) \geqq p_{ik}(s) \geqq [e^{-q_i(t-s)} - 1]\, p_{ik}(s)$$

$$\geqq -q_i(t-s)\, p_{ik}(s) .$$

[1] These statements generally aren't true for s = 0. More information is given in Proposition 7.

Take $0 \leq u < v$, cut $[u, v]$ into small pieces and apply this inequality to each small piece. There results

$$(1) \qquad p_{ik}(v) - p_{ik}(u) \geqq - \int_u^v q_i \, p_{ik}(s) \, ds \quad .$$

Now set

$$(2) \qquad g_k(t) = p_{ik}(t) - \delta_{ik} + \int_0^t q_i \, p_{ik}(u) du \quad .$$

This is positive according to (1). Let $G(t)$ be the row vector with coefficients $g_k(t)$: a routine calculation shows that $G(t+s) = G(s) + G(t)\Pi(s)$. According to Proposition 2, G' exists, is finite and continuous in T^0, and has a finite limit $G'(0+)$ at 0: a well known calculus theorem then implies that $G'(0+)$ can be interpreted as the right derivative of G at 0. This proves statement 1). Proposition 2 also asserts that G' is an entrance law, which is equivalent to statement 2).

REMARK. All that has been done till now is valid for GTM's (stochasticity has not been used). This will be applied later on (Proposition 8).

We now proceed to prove 3) and 4). The relation $G(t+s) - G(s) = G(t)\Pi(s)$ and the positivity of G imply that G increases with t. We get by differentiation:

$$(3) \qquad p'_{ik}(t) + q_i \, p_{ik}(t) \geqq 0 \quad .$$

We may assume that $q_i > 0$ (everything is trivial if i is absorbing). Let us set

$$(4) \qquad r_{ij}(t) = \frac{1}{q_i} [p'_{ij}(t) + q_i \, p_{ij}(t)] \quad .$$

It follows from (3) that $e^{q_i t} p_{ik}(t)$ increases with t. Since we can write the equation $\Sigma \, p_{ij}(t) = 1$ as $\Sigma \, p_{ij}(t) e^{q_i t} = e^{q_i t}$, Fubini's theorem on derivation of increasing functions gives us:

$$(5) \qquad \sum_j r_{ij}(t) = 1 \qquad \text{for almost all } t \, .$$

On the other hand, it is very easy to check that

(6)
$$r_{ij}(s+t) = \sum_k r_{ik}(s)\, p_{kj}(t)$$

and this implies at once that $\Sigma_j\, r_{ij}(s+t) = \Sigma_j\, r_{ij}(s)$. This sum therefore is a constant, and must be 1 from (5). So we get the first statement 3), and the second one follows at once by letting s tend to 0. Finally, we have

$$r_{ij}(t) - p_{ij}(t) = \frac{1}{q_i}\, p'_{ij}(t) \quad .$$

Therefore $q_i^{-1} |\, p'_{ij}(t)| \leqq r_{ij}(t) + p_{ij}(t)$. Summing over j, and applying the relations $\Sigma_j\, r_{ij} = \Sigma_j\, p_{ij} = 1$, we get 4).

PROPOSITION 7. *Assume that* i *is stable, and that* $\Sigma_{j \neq i}\, q_{ij} = q_i$. *Then the Kolmogorov "backward" differential equations hold at* i *:*

(1)
$$p'_{ij}(t) = \sum_k q_{ik}\, p_{kj}(t) \qquad (q_{ii} = -q_i)^{*} \quad .$$

Proof: We can write statement 2) of Proposition 6 as

$$p'_{ij}(s+t) + q_i\, p_{ij}(s+t) = \sum_k [p'_{ik}(s) + q_i\, p_{ik}(s)]\, p_{kj}(t) \quad .$$

Relation (3) of Proposition 6 implies that all terms are positive. Therefore we can let s tend to 0, and apply Fatou's lemma:

(2)
$$p'_{ij}(t) + q_i\, p_{ij}(t) \geqq \sum_k [q_{ik} + q_i \delta_{ik}]\, p_{kj}(t) \quad ;$$

deleting the term with q_i,

(3)
$$p'_{ij}(t) \geqq \sum_k q_{ik}\, p_{kj}(t) \quad .$$

---

* Conversely, it follows from 3) of Proposition 6 that if the Kolmogorov equations hold, then $\Sigma_{j \neq i}\, q_{ij} = q_i$.

If we sum these inequalities over j we get an equality. Therefore each of them must be an equality.

Finally, the following proposition isn't anything new: as we pointed out in the proof of Proposition 6, the first two statements are valid for GTM's. Therefore we can apply them to the transpose matrix function Π^*, getting the following result.

PROPOSITION 8. *Let j be a stable state. Then*

1) $p'_{ij}(t)$ *exists, and is finite and continuous for all* $t \geqq 0$ *and* $i \in I$;

2) $p'_{ij}(s+t) = \sum_k p_{ik}(s) p'_{kj}(t) \qquad (s \geqq 0,\ t > 0,\ j \in I)$.

In what follows, *all* states will be stable most of the time, and Proposition 8 will have very little interest.

CHAPTER II

SAMPLE FUNCTION BEHAVIOR

Organization. This chapter contains 3 sections: the first one consists of results which are true for all Markov chains (the lower semicontinuous version). Section 2 introduces one of our basic hypotheses, and deduces from it more precise results about sample functions (right continuity in $\mathbf{I}'$), and the strong Markov property. Section 3 is concerned with the first infinity, the minimal process, and the jump chain.

Section 1. Construction of Markov Chains

We start with the following result, which is a very special case of a general theorem on the construction of stochastic processes, and which will not be proved here.

PROPOSITION 1. *Let* Π *be a STM, and let* $p(\cdot)$ *be a normalized entrance law relative to* Π.[1] *Then it is possible to construct a complete probability space* $(\Omega, \underline{\underline{F}}, P)$. *and a stochastic process* $(X_t)_{t>0}$ *on* $(\Omega, \underline{\underline{F}})$ *with values in* $\mathbf{I}$, *such that* (X_t) *is a Markov process with* Π *as transition function, and* $p(\cdot)$ *as entrance law.*

Otherwise stated: if $0 < t_0 < t_1 < \cdots < t_n$ are arbitrary epochs, and $i_0, \ldots, i_n$ are states in $\mathbf{I}$, we have

$$P\{X(t_0) = i_0, X(t_1) = i_1 \cdots X(t_n) = i_n\}$$
$$= p_{i_0}(t_0) p_{i_0 i_1}(t_1 - t_0) \cdots p_{i_{n-1} i_n}(t_n - t_{n-1}) \; .$$

[1] If $p(t) = p\Pi(t)$ for some probability law p on $\mathbf{I}$, the process can be defined on $\mathbf{T}$ instead of $\mathbf{T}^0$, and the results below are valid on $\mathbf{T}$.

We shall use freely the notations X_t, $X_t(\omega)$, $X(t)$, $X(t,\omega)$ in the sequel.

NOTATION. We denote by $\underline{\underline{H}}_t$ $(t > 0)$ the σ-field generated by all random variables $X(s)$, $s \leqq t$, and all sets of measure 0.

PROPOSITION 2. *The chain* $X(t)$ *is stochastically continuous in the discrete topology. Otherwise stated: for all* $t > 0$

(1) $$P\{X(t+h) \neq X(t)]\} \underset{h \to 0+}{\to} 0 \quad ;$$

(2) $$P\{X(t-h) \neq X(t)\} \underset{h \to 0+}{\to} 0 \quad .$$

Proof: The first member of (1) is $\Sigma\, p_i(t)[1-p_{ii}(h)]$. It tends to 0 with h from Lebesgue's theorem. Let $u_h = P\{X(t-h) = X(t)\}$. We have

$$u_h = \sum_i p_i(t-h)p_{ii}(h) \leqq 1 \quad .$$

Now as h tends to 0, $p_i(t-h)$ tends to $p_i(t)$ and $p_{ii}(h)$ to 1, therefore we get from Fatou's lemma

$$1 = \sum_i p_i(t) \leqq \liminf_{h \to 0+} u_h$$

and the proposition follows.

The following (easy) result is technically very useful for work with optional times.

PROPOSITION 3. *The family* $(\underline{\underline{H}}_t)$ *is a right continuous family of* σ*-fields*.

Proof: Let t be any number > 0. We are going to prove that

(1) $$E[F \mid \underline{\underline{H}}_{t+}] = E[F \mid \underline{\underline{H}}_t] \qquad \text{a.s.}$$

for any bounded random variable F measurable with respect to $\underline{\underline{H}}_\infty$. Applying this to $F = I_A$, $A \in \underline{\underline{H}}_{t+}$, will give the proposition. According to well known theorems, it is sufficient to prove (1) when

$$F(\omega) = F_1 \circ X_{s_1}(\omega) \cdots F_k \circ X_{s_k}(\omega)$$

with $s_1 < \cdots < s_k$, $F_1, \ldots, F_k$ being bounded functions on $\mathbf{I}$. We may assume that $t < s_1$, because all factors with $s_i \leqq t$ can be taken out of both expectation signs. Also we may first take conditional expectations with respect to $\underline{\underline{H}}_{s_1}$, on both sides of (1), and we are thus reduced to proving that

$$E\left[G \circ X_{s_1} \mid \underline{\underline{H}}_{t+}\right] = E\left[G \circ X_{s_1} \mid \underline{\underline{H}}_t\right] \quad \text{a.s.} \tag{2}$$

if G is a bounded function on $\mathbf{I}$. And now we can assume that G is just the indicator function of some state j.

On the other hand, let t_n be a strictly decreasing sequence of elements of $]t, s_1[$ that converges to t. The first member of (2) is equal to $\lim_n E[G \circ X_{s_1} \mid \underline{\underline{H}}_{t_n}]$ a.s., according to elementary martingale theory. So we are finally reduced to proving that

$$E\left[I_{\{j\}} \circ X_{s_1} \mid \underline{\underline{H}}_{t_n}\right] \to E\left[I_{\{j\}} \circ X_{s_1} \mid \underline{\underline{H}}_t\right] \quad \text{a.s.} \tag{3}$$

This can be explicitly written (using the notation $p(t, i, j)$ for $p_{ij}(t)$) as

$$\lim p(s_1 - t_n, X_{t_n}, j) = p(s_1 - t, X_t, j) \ . \tag{4}$$

Now the limit in the first member *exists*, and so we may replace the sequence t_n by any subsequence t_n'. On the other hand, X_{t_n} converges in probability to X_t in the discrete topology (Proposition 1), and therefore we can choose the subsequence t_n' such that $X_{t_n'}$ converges *a.s.* to X_t in the discrete topology (this is well known for real valued random variables, but we can take $\mathbf{I}$ as the integers, in order to apply it here). This means that for a.e. ω, $X_{t_n'}(\omega) = X_t(\omega)$ for n large enough, and then the relation

$$p(s_1 - t_n', X_{t_n'}(\omega), j) \to p(s_1 - t, X_t(\omega), j)$$

is deduced from the continuity of $p(\cdot, i, j)$ for all i, j. Q.E.D.

We shall now use some martingale theory, and need some notations. First of all, $\mathbf{S}$ *denotes a countable dense set in* $\mathbf{T}^0$ (this is very important, since S will be used in several statements below). For any rational $A > 0$, $j \in I$, let us consider the martingale on $]0, A]$

$$Y_{A,j}(s, \cdot) = P\{X_A = j \mid \underline{\underline{H}}_s\} = p(A-s, X_s(\cdot), j) \qquad (0 < s < A) .$$

According to Doob's theorem on the sample function behavior of martingales, there is a set $N_{A,j}$ of measure 0 such that, if $\omega \notin N_{A,j}$ the sample function $s \to Y_{A,j}(s, \omega)$ has right hand limits along S at all points of $[0, A[$, left hand limits along S at all points of $]0, A]$. Also, since the family $(\underline{\underline{H}}_t)$ is right continuous,* we have for all $t \in]0, A[$, denoting by $Y_{A,j}(t+, \cdot)$ the above right hand limit,

$$Y_{A,j}(t, \cdot) = Y_{A,j}(t+, \cdot) \qquad \text{a.s.}$$

But here the exceptional set $N_{A,j,t}$ on which this equality isn't true *depends on* t.

Set

$$N'_{A,j} = N_{A,j} \cup \bigcup_{t \in S} N_{A,j,t} ,$$

$$N = \bigcup_{\substack{A \text{ rational} \\ j \in I}} N'_{A,j} .$$

Since S is countable, N is a set of measure 0, and we can affirm that if $\omega \notin N$ the sample function $Y_{A,j}(\cdot, \omega)$ has right and left limits along S, and is right continuous along S at every $t \in S$.

We can now prove an important result, true for all STM's.

PROPOSITION 4. *There exists a set* M *of measure* 0 *such that, if* $\omega \notin M$

1) *The sample function* $X(\cdot, \omega)$ *has at most one right cluster value in* I *along* S, *and at most one left cluster value in* I *along* S *at every* $t \in T^0$ (but the sample function can have also the cluster value ∞ along

2) *For each* $t \in S$, $X(t, \omega)$ *is a right cluster value (belonging to* I *by definition of* X*) of* $X(\cdot, \omega)$ *along* $S - \{t\}$ (i.e., there is some subsequence

* Another argument can be given, using stochastic continuity in the discrete topology of I.

s_n of elements of S, strictly decreasing to t, such that $X(s_n, \omega) = X(t, \omega)$ for all n).

Proof: Consider the set N of measure 0 defined above, and assume $\omega \notin N$. Let $t \in T^0$, and s_n, s'_n be two sequences of elements of S, decreasing to t, such that $X(s_n, \omega) \to j \in I$, $X(s'_n, \omega) \to j' \in I$ (if $t \in S$, the sequence defined by $s_n = t$ for all n is allowed). We shall prove that this implies $j = j'$, a result slightly stronger than 1).

Take a rational number $A > t$, and a state $k \in I$. Then, according to the fact that $X(s_n, \omega) \to j$, $X(s'_n, \omega) \to j'$ in the *discrete* topology, we have

$$\lim_n Y_{A,k}(s_n, \omega) = \lim_n p(A - s_n, X_{s_n}(\omega), k) = p(A-t, j, k) \ ;$$

in the same way

$$\lim_n Y_{A,k}(s'_n, \omega) = p(A-t, j', k) \ .$$

According to the properties of N stated above, and the fact that $\omega \notin N$, we have $p(A-t, j, k) = p(A-t, j', k)$. Letting A tend to t, we get $\delta_{jk} = \delta_{j'k}$, and therefore $j = j'$. The proof for the left cluster value is the same.

To prove 2), consider $t \in S$, and choose a sequence t_n strictly decreasing to t such that $t_n \in S$ and $X_{t_n} \to X_t$ a.s. in the discrete topology (Proposition 2), and denote by N'_t the set $\{X_{t_n} \nrightarrow X_t\}$. Since S is countable, $N' = \cup_{t \in S} N'_t$ is a set of measure 0, and we just set $M = N \cup N'$.

DEFINITION 1. M *being defined as in Proposition 4, put on* I *any ordering isomorphic to that of the positive integers, identify* ∞ *with* $+\infty$, *and set*

$$X'_t(\omega) = \infty \qquad \textit{for all } t > 0 \textit{ if } \omega \in M \ ;$$

$$X'_t(\omega) = \liminf_{s \to t,\, s \in S,\, s > t} X_s(\omega) \quad \textit{if } \omega \notin M, \textit{ for all } t > 0 \ .$$

Then the process (X'_t) *with values in* I' *is called the right lower semi-continuous version of the chain (* rlsc *version).*

This definition requires some comments. First of all, the order on I is entirely immaterial (its only use is compactness in writing and also the fact that it shows trivially that X_t' is a r.v.): indeed, according to Proposition 4, the lim inf is the only cluster value of $X(\cdot, \omega)$ at t along S, provided such a cluster value is available. If no such cluster value is available, then the lim inf is ∞. This equivalent definition is independent of the ordering.

PROPOSITION 5. *The rlsc version* (X_t') *is a standard modification of* (X_t) (i.e., $X_t = X_t'$ a.s. for each t, the exceptional set of measure 0 depending on t).

Proof: is the same as that of statement 2) of Proposition 4.

Note that Proposition 5 justifies the word "version" in the expression *the rlsc version.*

PROPOSITION 6. *Let* (X_t') *be the rlsc version. Then, for all* ω *and* t, *we have*

$$X_t'(\omega) = \liminf_{\substack{s \to t \\ s > t}} X_s'(\omega) = \liminf_{\substack{s \to t \\ s \in S,\, s > t}} X_s'(\omega) .$$

At each $t \in T^0$, *the following situation holds: either* $X'(\cdot, \omega)$ *has no right cluster value in* I *(along* T^0, t *excluded), in which case* $X'(t, \omega) = \infty$; *ot it has one such cluster value* k *(and perhaps also the cluster value* ∞*), in which case* $X'(\cdot, \omega) = k$.

Proof: Everything is obvious if $\omega \in M$. On the other hand, we have $X_s'(\omega) = X_s(\omega)$ for all $s \in S$ if $\omega \notin M$ (Proposition 4), therefore the relation $X_t'(\omega) = \liminf_{s \to t,\, s \in S,\, s > t} X_s'(\omega)$ is just the definition of X' for $\omega \notin M$.

It is now very easy to show that, for every r, $\liminf_{t \to r,\, t > r} X_t'(\omega) = \liminf_{s \to r,\, s \in S,\, s > r} X_s'(\omega)$, and the first statement follows.

The second statement is proved as follows: let k be a finite cluster value along T^0 at t, and let t_n be a sequence decreasing to t such that $X'(t_n, \omega) = k$ for all n. Then (according to the first statement) we may find $t'_n \in S$, arbitrarily close to t_n, such that $X'(t'_n, \omega) = k$. Therefore k is also a cluster value along S—and in fact the only right cluster value at t along S (apply Proposition 4, and the fact that $X'_s(\omega) = X_s(\omega)$ if $s \in S$).

COROLLARY. *Let* S' *be another countable dense set, and let* X' *be the rlsc version constructed from* S' *instead of* S. *Then* X' *and* X'' *are undiscriminable: for a.e.* $\omega \in \Omega$ *we have* $X'(\cdot, \omega) = X''(\cdot, \omega)$ *identically.*

Proof: We may assume that $S \subset S'$ (if not, we compare both S and S' to $S \cup S'$). Also, we have for a.e. ω, $X_t(\omega) = X'_t(\omega)$ for all $t \in S'$, since S' is countable (Proposition 5), and therefore for a.e. ω

$$X''_t(\omega) = \liminf_{\substack{s \to t,\, s > t \\ s \in S'}} X'_t(\omega) \qquad \text{for all } t.$$

Now this lim inf is included between the two lim inf of Proposition 6, which are equal, and equal to $X'_t(\omega)$.

This corollary justifies the use of the words "rlsc version" without mentioning S. From now on, the rlsc version will be denoted simply by X instead of X'.

Section 2. First Basic Hypothesis and Consequences

All results in the preceding section were true for all STM's. In this section we shall get much more precise results as consequences of the following hypothesis, which will be assumed in all the sequel.

FIRST BASIC HYPOTHESIS. *For every state* i, *we have* $q_i < \infty$ *and* $\Sigma_{j \neq i}\, q_{ij} = q_i$ (we assume always that Π is a STM, but we shall point out the extensions to the substochastic case).

Otherwise stated, all states are stable, and the Kolmogorov backward equations are satisfied (Chapter I, Proposition 7).

Then let us construct the rlsc version as in Section 1: we denote it by (X_t). Our first step will consist in proving the following theorem; we use in the proof only the fact that all states are stable.

THEOREM 1. *For almost every* ω, *the sample function* $X(\cdot,\omega)$ *(taking values in the Alexandrov compactification* $\mathbf{I}'$*) is right continuous and has a left hand limit at every* $t \in T^0$.[1]

NOTE. If Π is substochastic we have the same result, with just one absorbing state added to $\mathbf{I}$.

The proof of this theorem will require some notations and lemmas.

NOTATION. *For every state* i *and* $\omega \in \Omega$:

$$S_i(\omega) = \{t \in T^0 : X(t,\omega) = i\} ,$$

$$S_\infty(\omega) = \{t \in T^0 : X(t,\omega) = \infty\} ,$$

$$S_{\mathbf{I}}(\omega) = \bigcup_{i \in \mathbf{I}} S_i(\omega) .$$

The connected components of $S_i(\omega)$ are intervals (we shall see later on that these intervals are of the type [[): we shall call them i-intervals in the sequel.

LEMMA 1. $P\{X(t) = i \text{ for all } t \in [s, s+h] \mid X(s) = i\} = e^{-q_i h}$.

Proof: Let D be the set of all numbers $s + h(k/2^n)$ (n integer, k integer $\leqq 2^n$). We shall first prove that

$$\{X(t) = i \text{ for all } t \in [s,s+h]\} = \{X(t) = i \text{ for all } t \in D\} .$$

Indeed, the first set is contained in the second one; on the other hand, if ω belongs to the second set, then $X(\cdot,\omega)$ is equal to i for $t = s + h$, and has the right cluster value i along T^0 at every $t \in [s, s+h[$ —therefore $X(\cdot,\omega) = i$ on $[s, s+h]$ according to Proposition 6.

[1] As in Section 1, if the entrance law p(t) is such that $p(t) = p\Pi(t)$ for some probability law p, the results (here, right continuity) hold for $t = 0$.

The end of the proof is very easy: the probability we are looking for is equal to the limit of $P\{X(t) = i \text{ for } t = s + h(k/2^n),\ k = 1, \dots, 2^n \mid X(s) = i\}$, i.e., to $\lim_n (p_{ii}(h2^{-n}))^{2^n} = e^{-q_i h}$.

LEMMA 2. *Let* t *be fixed. Then, for a.e.* $\omega \in \Omega$ *such that* $t \in S_i(\omega)$, *there exists an open interval containing* t *and contained in* $S_i(\omega)$.

Proof: Let ε be a number > 0. Then

$$P\{X(s) = i \text{ for } s \in [t-\varepsilon, t+\varepsilon] \mid X(t) = i\} = \frac{p_i(t-\varepsilon)e^{-2q_i\varepsilon}}{p_i(t)}$$

and this tends to 1 as $\varepsilon \to 0$.

LEMMA 3. *Let* A *and* B *be two rational numbers such that* $0 < A < B < \infty$. *Then for a.e.* $\omega \in \Omega$ *the set* $S_i(\omega) \cap [A, B]$ *has a finite number of connected components.*

Proof: Let us denote by $D(D_n)$ the set of all numbers $A + (B-A)k/2^n$ for all n (fixed n) and all $k \leqq 2^n$. We shall use the following result, which is obvious and has nothing to do with probability:

"Let U be any subset of $[A, B]$ which possesses at least $m+1$ components. Then one may find numbers s_j, t_j $(1 \leqq j \leqq m)$ such that

$$A \leqq s_1 < t_1 < s_2 < t_2 < \cdots < s_m < t_m \leqq B ,$$

each s_j belongs to U and each t_j to U^c."

Denote by H_m the set of all ω such that $S_i(\omega) \cap [A, B]$ has at least $m+1$ components. Then we may find numbers s_j, t_j as above. On the other hand, according to the corollary to Proposition 6, we may assume that our basic countable dense set S, in the definition of the rlsc version, is such that $S \cap [A, B] = D$. Then it is obvious from the construction of the version that the s_i, t_i can be chosen *belonging to* D, and they belong therefore to some D_n. Otherwise stated,

$$P(H_m) \leqq \lim_{n\to\infty} P\{\text{there exist } s_1 < t_1 < \cdots < s_m < t_m \text{ in } D_n \text{ such that}$$
$$X(s_j) = i,\ X(t_j) \neq i \text{ for every } j = 1, \ldots, m\} .$$

Denote by H_{mn} the set on the right. Our lemma will be proved if we can show that

$$\lim_{m\to\infty} \lim_{n\to\infty} P(H_{mn}) = 0 .$$

Consider new the discrete Markov chain $Y(\nu) = X(A + \nu \frac{B-A}{2^n})$, set $M = 2^n$, and apply the following elementary lemma, which follows at once from the discrete strong Markov property:

Assume that for every $k \leqq M$, $P_i\{Y(\nu) \neq i \text{ for } 0 < \nu < k,\ Y(k) = i\} \leqq \lambda$ *Then* $P\{$there exist integers $s_1 < t_1 < \cdots < s_m < t_m \leqq M$ such that for all j, $Y(s_j) = i,\ Y(t_j) \neq i\}$ *is at most* λ^{m-1}.

Here we have $P_i\{Y(\nu) \neq i \text{ for } 0 < \nu < k\} \leqq P_i\{X(s)$ isn't identically equal to i on $[0, B-A]\} = 1 - \exp(-q_i(B-A))$. Denote this number by λ, and use in the lemma this λ *that doesn't depend* on n. Therefore, with this value of λ we have $\lim_{n\to\infty} P(H_{mn}) \leqq \lambda^{m-1}$, and we get the lemma as $m \to \infty$.

LEMMA 4. *For a.e.* $\omega \in \Omega$, *all* i*-components are half open intervals of the type* [[.[2]

Proof: We know that $S_i(\omega)$ a.s. has a finite number of components in each compact interval of T^0. These components, as connected subsets of the line, are intervals of the type [[, [ ],]] or] [. We must rule out the three last types. This is easily done for the last two, because of Proposition 6 (the relation $X(s, \omega) = i$ for $s \in\ ]a, b[$ would imply $X(a, \omega) = i$ by right lsc). As for type 2 it is ruled out by the fact that no point in $S_i(\omega)$ is isolated from the right ($X'_t(\omega) = \liminf X'_s(\omega)$ as $s \to t$ by *strictly* greater values: Proposition 6).

[1] P_i denotes a probability starting at i (for the discrete chain). Same notations 3 lines below, for discrete and continuous chains.

[2] For an open chain we may have an interval of the form $]0, t[$.

END OF THE PROOF OF THEOREM 1. Since $S_i(\omega)$ a.s. is a union of half open intervals of the type [[, the sample function $X(\cdot,\omega)$ a.s. is right continuous in $\mathbf{I}'$ at each point of $S_i(\omega)$. Since there are countably many $i \in \mathbf{I}$, we have a.e. right continuity at every point of $S_{\mathbf{I}}(\omega)$. On the other hand, right lower semicontinuity implies right continuity at every point of $S_\infty(\omega)$.

Let us prove now that left limits exist. Indeed, assume $X(\cdot,\omega)$ has no left limit at $t > 0$: then there must be at least two cluster values of $X(s,\omega)$ as $s \uparrow t$, and at least one of them (which we call i) must belong to $\mathbf{I}$: then $S_i(\omega)$ must have an infinity of components in every interval $[t-\varepsilon, t]$, contradicting Lemma 3. Theorem 1 is proved.

The following statement is a useful complement to Theorem 1.

PROPOSITION 7. *For a.e.* ω, *the following properties hold:*

1) $S_{\mathbf{I}}(\omega)$ *is the union of a countable number of disjoint intervals of the types* [[*or*] [*(we shall call them* $\mathbf{I}$*-intervals*).*

2) $S_\infty(\omega)$ *is a right closed set with Lebesgue measure* 0 *(and therefore empty interior).*

Proof: $\{\infty\}$ is a closed set in $\mathbf{I}'$, and the sample function is right continuous, therefore $S_\infty(\omega)$ is right closed, $S_{\mathbf{I}}(\omega)$ right open.[1] Property 1) is true for all right open sets in T^0 (their connected components are intervals of the above types). To prove statement 2, we need only prove (according to Fubini's theorem) that

$$E\left[\int_0^\infty I_\infty \circ X_t \, dt\right] = \int_0^\infty P\{X(t) = \infty\}\, dt \qquad \text{is equal to } 0\,.$$

This is obvious, since $P\{X(t) = \infty\} = 0$ for all t.

---

* Thus, an $\mathbf{I}$-interval is an interval of t in which $X(t,\omega) \in \mathbf{I}$.

1 "Right closed" means "closed from the right", for instance an interval [[is right closed!

The hypothesis that $\Sigma_{j \neq i}\, q_{ij} = q_i$ will be used in the next proposition (last statement).

PROPOSITION 8. *For any* $t > 0$ * *set*

$$J(\omega) = \inf\{s > t :\ X(s,\omega) \neq X(t,\omega)\} \qquad \text{(first jump after } t\text{)}.$$

Then J *is an optional r.v., and we have*

$$P\{J > t+h \mid X(t) = i\} = e^{-q_i h} , \tag{1}$$

and (if $q_i > 0$)

$$P\{X(J) = j \mid X(t) = i\} = \frac{q_{ij}}{q_i} \qquad (j \neq i) . \tag{2}$$

Therefore, since $\Sigma_{j \neq i}\, q_{ij} = q_i$, *the process doesn't jump to* ∞.

Proof: (1) is Lemma 1 proved above. This implies in particular that if $q_i > 0$, J is a.s. finite. Since we have $X(t) \in \mathbf{I}$ a.s. for every $t > 0$, we can (by considering the chain $Y(s) = X(t+s)$ instead of $X(s)$) reduce to the case $t = 0$. For any real number $r \geqq 0$, denote by $[r]$ the integral part of r, and set

$$J_m = 2^{-m}[2^m J + 1] ;$$

J_m takes only rational (dyadic) values and decreases to J as $m \to \infty$. We shall in fact prove a result stronger than (2). Set

$$A = \{J \geqq h,\ X(J) = j\} , \qquad A_m = \{J_m > h,\ X(J_m) = j\} .$$

Then $P\{A \mid X(0) = i\} = \lim_m P\{A_m \mid X(0) = i\}$. Now this probability can be computed as

---

* True also for $t = 0$ if $X(0+)$ exists and belongs to $\mathbf{I}$.

$$\sum_{n2^{-m}>h} P_i\{J_m = n2^{-m}, X(n2^{-m}) = j\} = \sum_{n2^{-m}>h} \exp\left[\frac{-(n-1)q_i}{2^m}\right] p_{ij}(2^{-m})$$

$$= \frac{\exp\{-[2^m h]q_i/2^m\}}{1-e^{-q_i/2^m}} q_{ij}(2^{-m} + o(2^{-m}))$$

and the limit is equal to $e^{-q_i h}(q_{ij}/q_i) = P\{J \geqq h\}(q_{ij}/q_i)$. Therefore we get

$$P\{X(J) = j \mid X(0) = i, J \geqq h\} = \frac{q_{ij}}{q_i} \quad \text{for every } h .$$

THE STRONG MARKOV PROPERTY. We are going to prove now a version of the strong Markov property. This version isn't quite sufficient, because we cannot compute the entrance law of the post T process if we haven't $X_T \in \mathbf{I}$ a.s. We shall see later on how it can be improved.

NOTATIONS. $(\Omega, \underline{\underline{F}}, P)$ is a complete probability space, $(X_t)_{t \in T^0}$ (or $t \in T$) is a right continuous Markov chain, with Π as transition matrix, and relative to a family $(\underline{\underline{H}}_t)$ of σ-fields which is right continuous. We assume also that each $\underline{\underline{H}}_t$ contains all P-null sets for convenience, but this is inessential.

It is also very convenient to assume that some absorbing point ∂ has been adjoined to $\mathbf{I}'$, and to set $X_\infty = \partial$. This allows the definition of the post-T process on the set $\{T = \infty\}$, and spares some boring discussions.

PROPOSITION 8. *Let* T *be an optional r.v. (with respect ot the family* $(\underline{\underline{H}}_t)$*).*

1) *The post-*T *process (with values in* $\mathbf{I}'$*)*

$$Y(t) = X(T+t)_{(t>0)}$$

is a right continuous Markov chain, with the transition matrix function Π. *This implies in particular that* $P\{X_{T+t} = \infty\} = 0$ *for each* $t > 0$.

2) *Assume that* $X_T \in \mathbf{I}$ *a.s. on* $\{T < \infty\}$ *(and that* $T > 0$ *a.s. if the parameter set of* X *is* T^0*). Then the entrance law of the post-*T *process*

is given by:

$$P\{Y(t) = i\} = \sum_j P\{X(T) = j, T < \infty\} p_{ji}(t) \qquad (i \in I),$$
$$P\{Y(t) = \partial\} = P\{T = \infty\} ;$$

(thus the value $t = 0$ can be added to the parameter set of the post-T process). *Moreover, the* σ *-field* $\underline{\underline{H}}'_T$ *generated by the post-* T *process is conditionally independent of* $\underline{\underline{H}}_T$ *given* X_T : *if* $L \in \underline{\underline{H}}_T$, $M \in \underline{\underline{H}}'_T$, $i \in I$,

$$\text{(1)} \qquad P\{L \cap M \mid X(T) = i\} = P\{L \mid X(T) = i\} P\{M \mid X(T) = i\} .$$ [1]

Proof: a) Let T be an optional r.v., let t_ν, $\nu = 1, \dots, n$ be strictly positive numbers in increasing order, j_ν, $\nu = 1, \dots, n$ be states in I. If we discard a set of measure 0, we may assume that $X(t) \in I$ for every rational t. Set $T_m = 2^{-m}[2^m T + 1]$ ([x] denotes here the greatest integer $\leqq x$, for every positive x): then T_m is a rational valued optional r.v. that decreases to T as $m \to \infty$. Every $L \in \underline{\underline{H}}_T$ belongs to $\underline{\underline{H}}_{T_m}$ and we have (since the strong Markov property is trivial for countably valued optional times)

$$P\{L, X(T_m) = i, X(T_m + t_\nu) = j_\nu \text{ for } \nu = 1, \dots, n\}$$
$$= P\{L, X(T_m) = i\} p(t_1, i, j_1) \prod_{\nu=1}^{n-1} p(t_{\nu+1} - t_\nu, j_\nu, j_{\nu+1}) .$$

Set now $M = \{X(T + t_\nu) = j_\nu, \ \nu = 1, \dots, n\}$, and let $T_m \downarrow T$. We get from the right continuity of the sample functions in the discrete topology:

$$P\{L, X(T) = i, M\} = P\{L, X(T) = i\} C ,$$

where C is $p(t_1, i, j_1) \prod_{\nu=1}^{n-1} p(t_{\nu+1} - t_\nu, j_\nu, j_{\nu+1})$. Taking $L = \Omega$, we get

[1] This equality tells nothing if $P\{X(T) = i\} = 0$. It can be extended to an optional r.v. T such that $P\{X(T) = i\} > 0$, but which does not satisfy the assumption $X(T) \in I$ a.s. on $\{T < \infty\}$ (the proof gives immediately this result

$C = P\{M \mid X(T) = i\}$ and formula (1) follows at once by a monotone class argument.

Assume that $X(T) \in I$ a.s., and let us show that the optional r.v. $T+s$ has the same property. Indeed, from the above formula we have

$$\sum_{j \in I} p\{X(T+s) = j, X(T) = i\} = \sum_{j} P\{X(T) = i\} p(s,i,j) = P\{X(T) = i\} .$$

Summing over i we get that $P\{X(T+s) \in I\} = P\{X(T) \in I\} = P\{T < \infty\} = P\{T+s < \infty\}$. Since $X(T) \in I$ a.s. on $\{T < \infty\}$, the formula

$$P\{L, X(T) = i, X(T+t) = j\} = P\{L, X(T) = i\} p(t,i,j)$$

can be written

$$P\{X(T+t) \mid \underline{\underline{H}}_T\} = p(t, X(T), j) .$$

On the other hand, from the facts just proven above, we can replace T by $T+s$ in this relation. Otherwise stated, the post-T process is a Markov chain with Π as its transition matrix, and statement 2) is completely proved.

Let us prove statement 1): we need only show that $X(T+s) \in I$ a.s. on $\{T < \infty\}$, because then statement 2), applied to $T+s$, will give the result. Now

$$E\left[\int_0^\infty I_{\{\infty\}} \circ X(T+s) ds\right]$$
$$= E\left[\int_T^\infty I_{\{\infty\}} \circ X(r) dr\right] \leqq E\left[\int_0^\infty I_{\{\infty\}} \circ X(r) dr\right] = \int_0^\infty P\{X_r = \infty\} dr = 0 .$$

Therefore, we get from Fubini's theorem that $P\{X(T+s) = \infty\} = 0$ for almost every s (Lebesgue measure), and therefore for arbitrarily small values of s we have

$$X(T+s) \in I \quad \text{a.s. on } \{T < \infty\} .$$

According to the last paragraph, this implies the same property for all s, and Proposition 8 is proved.

Section 3. The First Infinity

CANONICAL PROCESS. It will be very convenient now to work with one explicitly defined chain, which will be called the canonical chain. The interest of this definition lies in the fact that all measures involved will be defined on the same sample space.

Let us denote by Ω the set of all functions ω from T^0 to I' which possess the nice properties we have considered in the preceding section:

– they are right continuous with left hand limits, and never jump from a state in I to ∞;

– the set $S_\infty(\omega)$ has measure 0 in the Lebesgue sense.

We shall denote by Ω_0 the set of those $\omega \in \Omega$ which have a finite right hand limit at 0 in the discrete topology. If $\omega \in \Omega$, we set $X_t(\omega)$ (or $X(t,\omega)$) $= \omega(t)$, and if $\omega \in \Omega_0$ we denote the limit at 0 by $X_0(\omega)$ (or $X(0,\omega)$). We also define $X_{t-}(\omega)$ (or $X(t-,\omega)$) $= \omega(t-)$.

For every normalized entrance law $p(\cdot)$ there is one (and only one) law P on Ω (given the σ-field generated by all the r.v.'s $X(t)$, $t > 0$) such that $X(t)$ is a Markov process with the transition semi-group $\Pi(\cdot)$ and the entrance law $p(\cdot)$. P is concentrated on Ω_0 iff $p(t) = \mu\Pi(t)$ for some probability law μ on I, and then μ is just the law of $X(0)$. We shall not use a notation emphasizing the dependence of P on $p(\cdot)$, except in a few cases, the most noteworthy of which are:

1) If $p(t) = \mu\Pi(t)$ for some law μ on I, we shall sometimes write P_μ instead of P. In particular, if μ is the unit mass at $i \in I$, we shall write P as P_i (in this case $p_j(t) = p_{ij}(t)$ for all j and t).

2) Cases to be seen later, of processes starting at a boundary point.

To every law P we can associate natural completed σ-fields $\underline{\underline{H}}_t$ (see Proposition 3). If $P = P_\mu$, we shall write sometimes $\underline{\underline{H}}_t^\mu$ to emphasize the dependence on the completion of μ – only if necessary.

The use of canonical process allows that of Dynkin's translation operator θ_t (defined by $X(s,\theta_t\omega) = X(s+t,\omega)$), which is sometimes quite convenient. For instance, the strong Markov property can be written as follows if T is an optional r.v. such that $X(T) \in I$ a.s. on $\{T < \infty\}$, then

$$P(\theta_T^{-1}A \mid \underline{\underline{H}}_T) = P_{X(T)}(A) \qquad \text{a.s. for every } A \in \underline{\underline{H}}_\infty \ .$$

We shall define now some basic optional times:

1. SUCCESSIVE JUMPING TIMES. We set for every $\omega \in \Omega_0$:
   a) $J_0(\omega) = 0$, $\chi_0(\omega) = X_0(\omega)$;
   b) $J_1(\omega) = \inf\{s > 0 : X(s,\omega) \neq X(s-,\omega)\}$ (first jump time), $\chi_1(\omega) = X(J_1(\omega),\omega)$ (first jump value: it belongs to I, since there is no jumping to ∞) .
   c) We now set inductively
   $J_n(\omega) = \inf\{s > J_{n-1}(\omega) : X(s,\omega) \neq X(s-,\omega)\}$ (n-th jump time),
   $\chi_n(\omega) = X(J_n(\omega),\omega)$.

These definitions aren't entirely satisfactory, because there may be some absorbing states, and it is slightly invonvenient to exclude them. The presence of absorbing states will reveal itself in the fact that some J_n will have the value $+\infty$, and the corresponding χ_n will be undefined. Therefore, we shall make the convention that if n is the last index such that $J_n(\omega) < \infty$, then $\chi_m(\omega) = \chi_n(\omega)$ for all $m > n$.

2. THE FIRST INFINITY. We set for every $\omega \in \Omega$ *

$$\tau(\omega) = \inf\{s > 0 : X(s-,\omega) = \infty\} \quad .$$

If $\omega \in \Omega_0$, we have $\tau(\omega) > 0$ from right continuity, and obviously $\tau(\omega) = \lim J_n(\omega)$: τ is the "first infinity" (note that ∞ is never hit before time τ, since there are no jumps to infinity).

Proposition 7 and the strong Markov property give us at once the following proposition:

---

* In the substochastic case, we add an absorbing state ∂ to I to obtain true probabilities. It is convenient then to set $\tau(\omega) = \inf\{s > 0 : X(s,\omega) \in \{\infty, \partial\}\}$. It follows from Proposition 9 that χ_n does not take the value ∂.

PROPOSITION 9. *Let* μ *be a probability law on* $\mathbf{I}$. *Then the process* (χ_n) *is, for the law* P_μ, *a discrete parameter Markov chain with the initial law* μ *and the transition matrix* (s_{ij}) *given by*

a) *if* i *is nonabsorbing,* $s_{ii} = 0$ *and* $s_{ij} = (q_{ij}/q_i)$ *if* $j \neq i$;

b) *if* i *is absorbing,* $s_{ij} = \delta_{ij}$.

We shall say that (χ_n) is the imbedded *jump chain.*

3. THE MINIMAL SEMI-GROUP. We shall denote by $\Phi(t)$ the (substochastic) matrix on $\mathbf{I} \times \mathbf{I}$ with the following coefficients

$$f_{ij}(t) = P_i\{X(t) = j,\ t < \tau\} ,$$

and by $\tilde{\Phi}(t)$ the stochastic matrix on $(\mathbf{I} \cup \{\partial\}) \times (\mathbf{I} \cup \{\partial\})$ * with the same coefficients $f_{ij}(t)$ as above, and such that

$$f_{i\partial}(t) = P_i\{t \geqq \tau\} = 1 - \sum_j f_{ij}(t) \qquad i \in \mathbf{I} \quad ,$$

$$f_{\partial\partial}(t) = 1, \quad f_{\partial i}(t) = 0 \qquad \text{for} \qquad i \in \mathbf{I} .$$

We shall not prove the following proposition (which is easy).

PROPOSITION 10. 1) $\Phi(\cdot)$ *is a* SSTM *on* $\mathbf{I}$, *and* $\tilde{\Phi}(\cdot)$ *a* STM *on* $\mathbf{I} \cup \{\partial\}$. 2) *Set* $Y_t(\omega) = X_t(\omega)$ *if* $t < \tau(\omega)$, $Y_t(\omega) = \partial$ *for* $t \geqq \tau(\omega)$. *Then the process* Y_t, *if* Ω *is given the measure* P_μ, *is a Markov chain which has* $\tilde{\Phi}$ *as transition matrix function,* μ *as initial law.*

It is obvious from the probabilistic interpretation that the two processes Y and X have the same jump chain, and therefore the same coefficients q_i and q_{ij}. On the other hand, we have $f_{ij}(t) \leqq p_{ij}(t)$ for all t, i, j (on $\mathbf{I}$). It can be proved (easily, but we shan't do it here) that the coefficients $f_{ij}(\cdot)$ depend only on the coefficients q_i and q_{ij}, and that $\Phi(\cdot)$ is the *smallest* SSTM on $\mathbf{I}$ which admits the q_i, q_{ij} as derivatives at 0. This is the reason why it is called the *minimal* semi-group.

---

* As usual, ∂ is the extra point added to have true probability measures.

We are going now to define the "exits to infinity", according to Blackwell. This approach isn't entirely satisfactory in the general case, but since we shall restrict ourselves to the case of an atomic (even finite) boundary, it will be quite sufficient for our purposes.

NOTATION. Let A and B be two subsets of Ω_0. We write $A \doteq B$ to mean that the symmetric difference $A \,\Delta\, B$ is a set of probability 0 for every law P_μ. Let us choose one fixed probability measure η on $\mathbf{I}$, such that $\eta(i) > 0$ for all i; then $A \doteq B$ iff $P_\eta(A \,\Delta\, B) = 0$. We say that A is a *null set* iff $A \doteq \emptyset$.

The strong Markov property implies at once that, if T is an optional r.v. such that $\{X(T) = \infty\} \doteq \emptyset$ (we shall also assume, for simplicity, that $\{T = \infty\} \doteq \emptyset$), and if $A \doteq B$, then we also have the relation $\theta_T^{-1} A \doteq \theta_T^{-1} B$ between their translates.

DEFINITIONS ON $\mathbf{I}$. A subset A of $\mathbf{I}$ is *transient* (for the jump chain) if $\overline{\lim}\{\chi_n \in A\} \doteq \emptyset$ —otherwise stated iff, for P_η a.e. ω, there are only a finite number of jumps into A.

A subset B of $\mathbf{I}$ is *almost closed* (for the jump chain) if $\overline{\lim}\{\chi_n \in B\} \doteq \underline{\lim}\{\chi_n \in B\}$ —otherwise stated, iff a.e. ω, which meets B infinitely often before time τ, in fact, remains ultimately in B.

We shall denote by $\underline{\underline{C}}$ the set of all almost closed sets, and by $\underline{\underline{T}}$ the set of all transient subsets of $\mathbf{I}$. We then have the following result.

PROPOSITION 11. $\underline{\underline{C}}$ *is a Boolean algebra* (i.e., contains $\emptyset$, $\mathbf{I}$, is closed under *finite* unions and intersections and complementation), *and* $\underline{\underline{T}}$ *is an ideal in* $\underline{\underline{C}}$ (i.e., $\underline{\underline{T}} \subset \underline{\underline{C}}$ and if $A \in \underline{\underline{T}}$ then every $B \subset A$ belongs to $\underline{\underline{T}}$).

Proof: We have

$$\overline{\lim}\{\chi_n \in A^c\} = \Omega \setminus \underline{\lim}\{\chi_n \in A\},\ \underline{\lim}\{\chi_n \in A^c\} = \Omega \setminus \overline{\lim}\{\chi_n \in A\}\ .$$

Therefore, $\underline{\underline{C}}$ is closed under ^{c}. To prove that it is closed under $\cup$, we write

$$(\overline{\lim}\setminus\underline{\lim})\{\chi_n \in A \cup B\} \subset (\overline{\lim}\setminus\underline{\lim})\{\chi_n \in A\} \cup (\overline{\lim}\setminus\underline{\lim})\{\chi_n \in B\} .$$

The remainder of the proposition is trivial.

The interest in the Boolean algebra $\underline{\underline{C}}$ comes from the following fact: take a finite partition of I into almost closed sets $A_1, \ldots, A_n$. Then for every $\omega \in \Omega$ we can find some k such that $\chi_n(\omega) \in A_k$ for infinitely many n. Therefore (from almost closedness) we can assert that

> for a.e. ω there is one and only one k such that $\chi_n(\omega) \in A_k$ for *all* sufficiently large n.

Hence the sets $A_1, \ldots, A_m$ describe mutually exclusive possibilities of the behavior of the jump chain as $n \to \infty$. On the other hand, if A and B differ by a transient set, a.e. sample function that stays in A for n large enough stays also in B for n large enough, and hence A and B describe the same behavior as $n \to \infty$: this is why the interesting algebra is $\underline{\underline{C}}/\underline{\underline{T}}$ rather than $\underline{\underline{C}}$.

DEFINITIONS ON Ω_0. $\underline{\underline{G}}^0$ is the σ-field generated by the random variables $\chi_0, \chi_1, \ldots$ of the jump chain, and $\underline{\underline{G}}$ is the σ-field consisting of those $A \subset \Omega_0$ such that $A \doteq B$ for some $B \in \underline{\underline{G}}^0$ ($\underline{\underline{G}}$ is the completed σ-field of $\underline{\underline{G}}^0$ for P_η).

For any random variable f on Ω_0, denote by $T^n f$ the r.v. defined by $T^n f(\omega) = f(\theta_{J_n}\omega)$ (shifting by the n-th jump time). It is obvious that $T^n f = T(T^{n-1}f)$ (where $T = T^1$). We say that a r.v. f is invariant iff 1) it is $\underline{\underline{G}}$-measurable, 2) $f \doteq Tf$; obviously f is invariant iff it is measurable with respect to the σ-field of *invariant* subsets of Ω_0. We denote this σ-field by $\underline{\underline{I}} \subset \underline{\underline{G}}$. We also denote by $\underline{\underline{N}}$ the family of all null sets in Ω_0.

In the following proposition, we denote by $S = (s_{ij})$ the transition matrix of the jump chain.

PROPOSITION 12. *Let* f *be the indicator of an invariant set* H, *and let* h *be the function on* I *defined by*

$$h(i) = P_i(H) \quad .$$

1) *h is an invariant function for the jump matrix:* $Sh = h$.

2) *For any measure* P_μ, *the process* $h \circ \chi_n$ *is a martingale, and we have*

$$f \doteq \underline{\lim}\, h \circ \chi_n \doteq \overline{\lim}\, h \circ \chi_n \quad .$$

3) *Let* a *be any number such that* $0 < a < 1$. *Then the set* $A = \{h \geq a\}$ *is almost closed. If* a' *is another number in* $]0, 1[$, *the sets* $\{h \geq a\}$ *and* $\{h \geq a'\}$ *differ only by a transient set. We have* $H = \overline{\lim}\{\chi_n \in A\}$.

4) *The function* $H \to \{h \geq a\}$ *from* I *(the invariant sets in* Ω_0*) to* $\underline{\underline{C}}/\underline{\underline{T}}$ *(the almost closed sets in* I *modulo transient sets) is a Boolean algebra homomorphism, and induces an isomorphism from* $\underline{\underline{I}}/\underline{\underline{N}}$ *onto* $\underline{\underline{C}}/\underline{\underline{T}}$. (Warning: countable unions and intersections are defined in $\underline{\underline{I}}/\underline{\underline{N}}$, but not in $\underline{\underline{C}}/\underline{\underline{T}}$, and this isomorphism isn't compatible with infinite operations.)

Proof: The fact that $Sh = h$ is obtained by taking expectations with respect to P_i in the equality $Tf \doteq f$ (or $f \circ \theta_J \doteq f$). The fact that $(h \circ \chi_n)$ is a martingale is an obvious consequence of this result, or of the following more precise computation: since $Tf \doteq f$, we also have $f \doteq T^n f$ for every n, or

$$E[f \mid \underline{\underline{H}}_{J_n}] = E[T^n f \mid \underline{\underline{H}}_{J_n}] = E[f \circ \theta_{J_n} \mid \underline{\underline{H}}_{J_n}] = h \circ X(J_n) = h \circ \chi_n$$

(strong Markov property). According to the standard martingale convergence theorem, since f is a.s. equal to a r.v. measurable on the σ-field generated by all the χ_n's, we have

$$f = \lim h \circ \chi_n \qquad P_\mu\text{-a.s. for every law } \mu$$

which amounts to 2). Since f is an indicator function, the relation $\overline{\lim}\, h \circ \chi_n > 0$ a.s. implies $\lim h \circ \chi_n = 1$, which gives us at once the following properties:

$$\overline{\lim}\, h \circ \chi_n \geq a \text{ a.s. implies } \underline{\lim}\, h \circ \chi_n = 1,$$

and therefore $\underline{\lim}\, h \circ \chi_n \geqq a$. Therefore $A = \{h \geqq a\}$ is almost closed.

$f = 1$ a.s. implies $\underline{\lim}\, h \circ \chi_n \geqq a$, and conversely $\overline{\lim}\, h \circ \chi_n \geqq a$ a.s. implies $f > 0$, and hence $f = 1$. Therefore $H \doteq \overline{\lim}\{\chi_n \in A\}$.

If $a < a'$, $\overline{\lim}\, h \circ \chi_n \geqq$ a.s. implies $\underline{\lim}\, h \circ \chi_n \geqq a'$, therefore $\{h \geqq a\}$ and $\{h \geqq a'\}$ differ only by a transient set.

We leave 4) (which is now trivial) to the reader.

DEFINITION. *An exit (or boundary atom) is an atom of the Boolean algebra* $\underline{\underline{C}}/\underline{\underline{T}}$, *i.e., an almost closed set* A *which isn't transient, and is such that every almost closed* $B \subset A$ *either is transient, or differs from* A *by a transient set.*

PROPOSITION 3. I *can be decomposed into a sequence of disjoint almost closed sets*

$$I = I_0 \bigcup_{1 \leqq n < N} I_n$$

where $0 < N \leqq \infty$, *where each* I_n *for* $n \geqq 1$ *is an atom, and* I_0 *(the non-atomic part) contains no atom.*

Proof: The measure η having the same meaning as above, the σ-field I of invariant sets can be similarly decomposed (up to sets of P_η-measure 0, i.e., null sets). Then we come back to I in a trivial way.

We are going now to classify atoms according to whether they can be reached in finite time or not. We shall first need a lemma (which is Lévy's well-known generalization of the Borel-Cantelli lemma).

LEMMA. *Let* $(\Omega, \underline{\underline{G}}, P)$ *be any probability space, let* $\underline{\underline{G}}_n$ *be an increasing sequence of sub-*σ*-fields of* $\underline{\underline{G}}$, *and let* X_n *be a sequence of positive, uniformly bounded random variables such that* X_n *is* $\underline{\underline{G}}_n$*-measurable for every*

$n \geqq 0$. *Set* $X'_0 = X_0$, $X'_n = E[X_n \mid \underline{\underline{G}}_{n-1}]$ *for* $n > 0$. *Then*

$$\{\Sigma X_n = \infty\} \doteq \{\Sigma X'_n = \infty\} .$$

Proof: Set $Y_n = X_0 + \cdots + X_n$, $Y'_n = X'_0 + \cdots + X'_n$; the process $(Y_n - Y'_n)$ is a martingale.

a) Let us set $A' = \{\Sigma X'_n < \infty\}$, and prove that $\Sigma X_n < \infty$ a.s. on A' (boundedness will not be used here). Choose some $N > 0$, and set $T' = \inf\{n : Y'_{n+1} > N\}$: Since Y'_{n+1} is $\underline{\underline{G}}_n$-measurable for all n, T' is optional. Let Z' be the martingale $Y - Y'$ stopped at time T': then Z' is a martingale, and $Z' \geqq -N$. Therefore Z' converges a.s., and this implies $\Sigma X_n < \infty$ a.s. on $\{T' = \infty\}$. Letting $N \to \infty$ we get that $\Sigma X_n < \infty$ a.s. on A'.

b) Let $A = \{\Sigma X_n < \infty\}$, and $U = \sup_n X_n \in L^\infty$, $T = \inf\{n : Y_n \geqq N\}$. Let Z be the martingale $Y - Y'$ stopped at time T: then Z is a martingale smaller than $U + N \in L^\infty$, therefore it converges a.s. and this implies, as above, that $\Sigma X'_n < \infty$ a.s. on A. The lemma is proved.

Now we can prove the following proposition:

PROPOSITION 4. 1) *We have*

$$\{\omega \in \Omega_0 : \tau(\omega) < \infty\} \doteq \{\omega \in \Omega_0 : \sum_n \frac{1}{q_{\chi_n(\omega)}} < \infty\} .$$

2) *Let* A *be a boundary atom. Then either* $\lim\{\chi_n \in A\} \cap \{\tau = \infty\}$ *is a null set* (almost all sample functions going through A reach infinity in finite time: the atom is called a *passable* atom) *or* $\lim\{\chi_n \in A\} \cap \{\tau < \infty\}$ *is a null set* (it a.s. takes an infinite time to reach infinity through A: the atom is called *impassable*).

Proof: Denote by J_n the successive jumping times, and by $\underline{\underline{G}}_n$ the σ-field $\underline{\underline{H}}_{J_n}$. Set $x_0 = J_1 \wedge 1$, $x_n = (J_{n+1} - J_n) \wedge 1$; then $\{\tau < \infty\} = \{\Sigma x_n < \infty$. According to the lemma above, we have

$$\{\tau < \infty\} \doteq \{\Sigma E[x_n \mid \underline{\underline{H}}_{J_n}] < \infty\} .$$

Now the conditional law of $J_{n+1} - J_n$ given $\underline{\underline{H}}_{J_n}$ is an exponential distribution with the parameter q_{χ_n}. If x is an exponential r.v. with the parameter q, $E[x \wedge 1]$ is equal to $[1-e^{-q}]/q$. Therefore

$$\{\tau < \infty\} \doteq \left\{ \Sigma \frac{1}{q_{\chi_n}} \left[1 - e^{-q_{\chi_n}}\right] < \infty \right\} = \left\{ \Sigma \frac{1}{q_{\chi_n}} < \infty \right\} .$$

This implies that $\{\tau < \infty\}$ is an invariant set. If A is a boundary atom, we must then have either $\lim\{\chi_n \in A\} \subset \{\tau < \infty\}$. a.s., or the two sets have a null intersection.

We are now ready to state our fundamental hypothesis, which will be assumed from now on:

FUNDAMENTAL HYPOTHESIS 2. The invariant set $\{\tau < \infty\}$ is the union of a *finite* number of atomic invariant sets.

Otherwise stated, there is only a finite number of ways of going to infinity in finite time.

CHAPTER III

BOUNDARY BEHAVIOR

Section 1 contains the definitions of distinguishable exits, stickiness, and the basic strong Markov property at the boundary. Section 2 introduces the "boundary switching" and the quantities connected with it. Section 3 gives a number of lemmas on recurrence, necessary for the proof of Theorem 1, which is the main key to the structure of the process. The chapter concludes with the "complete decomposition formula". Some probabilistic interpretations will have to wait till Chapter IV.

Section 1. Crossing the Boundary

NOTATIONS. As in the preceding section (Chapter II), η denotes a probability measure on I such that every point has a strictly positive measure for η.

We denote by E the finite set of all *passable boundary atoms*. We can associate to each $a \in E$ a set $\Delta^a \subset \Omega$ such that

1) for each a, Δ^a belongs to $\underline{\underline{H}}_{\tau-}$ [1] and is invariant (note that $J_n \uparrow \tau$, and that $J_n < \tau$ for all n on $\{\tau < \infty\}$, [2] therefore $\underline{\underline{H}}_{\tau-}$ is just the σ-field $\bigvee_n \underline{\underline{H}}_{J_n}$ generated by all the σ-fields $\underline{\underline{H}}_{J_n}$),

2) the sets Δ^a are disjoint, and their union is $\{\tau < \infty\}$,

3) every invariant set A either contains a.s. Δ^a, or is a.s. disjoint from Δ^a.

[1] For the general definition of $H_{\tau-}$, see P. A. Meyer, "Guide détaillé de la théorie "générale" des processus". Séminaire de probabilités II, *Université de Strasbourg,* Springer-Verlag, 1968.

[2] Even on $\{\tau = \infty\}$ if there are no absorbing states.

We denote by τ^a the optional r.v. τ_{Δ^a}, i.e., $\tau^a = \tau$ on Δ^a, ∞ on $(\Delta^a)^c$.

All these random variables are defined on Ω_0. However, if we take any fixed $t > 0$, we have a.s. $X_t(\omega)$ $(= X_{t+}(\omega)) \in I$ (Chapter II, Proposition 5) and hence $\theta_t(\omega) \in \Omega_0$. The following random variables are therefore *a.e. defined on* Ω :

$$\tau_{(t)} = t + \tau \circ \theta_t \ , \qquad \tau^a_{(t)} = t + \tau^a \circ \theta_t \quad .$$

It will be convenient to define these random variables everywhere, that is, to define $\tau_{(t)}(\omega)$, $\tau^a_{(t)}(\omega)$ if $X_t(\omega) = \infty$. Since $S_\infty(\omega)$ has empty interior, we can define for instance

$$\tau^a_{(t)}(\omega) = \inf\{\tau^a_{(s)}(\omega), \text{ for all } s > t \text{ such that } X_s(\omega) \in I\} \ .$$

We know that the post-$\tau^a_{(t)}$ processes are Markov chains: their entrance laws will play a basic role in the sequel.

PROPOSITION 1. *Let* a *be an element of* E. *There is one and only one normalized entrance law* $(\xi^a_j(t))$ *relative to* Π *such that, for every initial law* μ

$$P_\mu\{X(\tau+t) = j, \Delta^a \mid \underline{\underline{H}}_{\tau-}\} = \xi^a_j(t) 1_{\Delta^a} \ , \qquad P_\mu\text{-a.s.}$$

Let P *be the law constructed on* Ω *from an arbitrary entrance law, and let* S *be an optional r.v. such that* $X(S) \in I$ *a.s. on* $\{S < \infty\}$. *Set*

$$\tau'^a = S + \tau^a \circ \theta_S \ , \qquad \tau' = S + \tau \circ \theta_S \ .$$

Then

$$P\{X(\tau' + t) = j, \tau'^a < \infty \mid \underline{\underline{H}}_{\tau'-}\} = \xi^a_j(t) 1_{\{\tau'^a < \infty\}} \ .$$ [1]

---

[1] Of course, in the substochastic case we have a coefficient $\xi^a_\partial(t)$.

Proof: Note that the first statement is just the particular case $S = 0$ of the second one.[2]

Let D be the event $\{X(\tau + t) = j, \Delta^a\}$, where $t > 0$, $j \in I$, $\partial \in E$. and set $h(i) = P_i(D)$. Set also $J'_n = S + J_n \circ \theta_S$ (n-th jump after S). The σ-field $\underline{\underline{H}}_{\tau'-}$ is generated by the increasing family of σ-fields $\underline{\underline{H}}_{J'_n}$. Therefore, using the standard martingale convergence theorem and the strong Markov property at time J'_n, we get that:

$$P\{X(\tau'+t) = j, \tau'^a < \infty \mid \underline{\underline{H}}_{\tau'-}\} \underset{\text{a.s.}}{=} \lim P\{X(\tau'+t) = j, \tau'^a < \infty \mid \underline{\underline{H}}_{J'_n}\}$$

$$= \lim P\{X(\tau'+t) = j, \tau'^a < \infty \mid X(J'_n)\} .$$

Now remark that $\{X(\tau'+t) = j, \tau'^a < \infty\}$ (the event that "the first hitting of ∞ after S occurs at a, etc."—a somewhat imprecise, but suggestive language) is just the translate of D by S, and is also the translate by J'_n: therefore, the strong Markov property implies that the last probability is just $h \circ X(J'_n)$. Summarizing, we have found that

$$P\{X(\tau'+t) = j, \tau'^a < \infty \mid \underline{\underline{H}}_{\tau'-}\} = \lim h \circ X(J'_n) \qquad \text{a.s.}$$

This is our first step. To identify this limit, let us turn to the case $S = 0$: then this relation can be written

$$P_\mu\{X(\tau+t) = j, \Delta^a \mid \underline{\underline{H}}_{\tau-}\} = \lim h \circ X(J_n) \qquad \text{a.s.}$$

Now the second member obviously is an invariant function, and therefore is a.s. constant on every atom. Taking first $\mu = \eta$, we find that the limit is equal P_η -a.s. to a constant C on Δ^a. Since $\eta(i) > 0$ for all i, we get that the limit is equal to $C\,P_\mu$-a.s. for every measure μ. On the other hand, $P_\mu\{X(\tau+t) = j, \Delta^a \mid \underline{\underline{H}}_{\tau-}\} \leq P_\mu\{\Delta^a \mid \underline{\underline{H}}_{\tau-}\} = 1_{\Delta^a}$, and therefore the limit is 0 on the complement of Δ^a. Therefore

---

[2] The stronger form will be needed for the proof of the strong Markov property at the boundary later on.

$$\lim h \circ X_{J_n} = P_\mu\{X(\tau+t) = j, \Delta^a \mid \underline{\underline{H}}_{\tau-}\} = C1_{\Delta^a} \qquad P_\mu\text{-a.s. for all } \mu.$$

This is our second step. The third step consists in applying the strong Markov property at time S to deduce from this the relation

$$\lim h \circ X_{J_n'} = P\{X(\tau'+t) = j, \tau'^a < \infty \mid \underline{\underline{H}}_{\tau'-}\} = C1_{\{\tau'^a < \infty\}} \quad \text{a.s.}$$

The only thing left now consists in identifying the constant C. To do it, take $\mu = \eta$ again (to be sure that Δ^a has a strictly positive probability), and denote by $z_j(t)$ the entrance law of the post-τ^a process for P_η. Then integrating the relation $P\{X(\tau+t) = j, \Delta^a \mid \underline{\underline{H}}_{\tau-}\} = C1_{\Delta^a}$ we get

$$z_j(t) = C P_\eta\{\tau^a < \infty\}$$

which proves that C is just the value of the *normalized* entrance law $z_j(t)/P_\eta\{\tau^a < \infty\}$. We set $C = \xi_j^a(t)$, and the proposition is proved.

Before we give some basic definitions and notations, we prove a lemma which gives a pleasant interpretation of $\xi_j^a(t)$. It can be proved that for a.e. ω the convergence is uniform on compact sets in T^0 (for fixed j) but this fact will not be needed.

PROPOSITION 2. *For almost all* $\omega \in \Delta^a$,

$$p_{X_n(\omega), j}(t) \underset{n\to\infty}{\to} \xi_j^a(t) \qquad \textit{for all } t > 0.$$

Proof: We shall first prove a.s. convergence for t fixed. We shall need a little lemma from martingale theory, due to Hunt.

LEMMA. *Let* $(W, \underline{\underline{G}}, P)$ *be any probability space, and let* $\underline{\underline{G}}_n$ *be an increasing family of sub σ-fields of* $\underline{\underline{G}}$. *Let* f_n *be any sequence of r.v., dominated by an integrable r.v., and converging a.e. to a r.v. f. Then*

$$E[f_n \mid \mathcal{G}_n] \to E[f \mid \bigvee_n \mathcal{G}_n] \qquad \text{a.s.}$$

Proof of the Lemma. For simplicity of notation, denote by $\underline{\underline{G}}'$ the σ-field $\vee_n \underline{\underline{G}}_n$ generated by the $\underline{\underline{G}}_n$'s. Since f_n converges to f a.s. we have for all H

$$\inf_{n>N} f_N \leq f \leq \sup_{n>N} f_n \qquad \text{a.s.}$$

Using the domination, we can find N so large that

$$E[\sup_{n>N} f_n - \inf_{n>N} f_n \mid \underline{\underline{G}}'] \leq \varepsilon \ .$$

Denote this sup by s, this inf by i. We have for all m and $n > N$

$$E[i \mid \underline{\underline{G}}_m] \leq E[f_n \mid \underline{\underline{G}}_m] \leq E[s \mid \underline{\underline{G}}_m]$$

Therefore, passing to the limit

$$E[i \mid \underline{\underline{G}}'] \leq \liminf E[f_m \mid \underline{\underline{G}}_m] \leq \limsup E[f_m \mid \underline{\underline{G}}_m] \leq E[s \mid G']$$

and of course $E[f \mid \underline{\underline{G}}']$ also lies between the left and right members. Since $|E[s \mid \underline{\underline{G}}'] - E[i \mid \underline{\underline{G}}']| \leq \varepsilon$, the result follows at once.

Let us prove Proposition 2 for fixed $t > 0$. We know that $J_n \uparrow \tau$, that $X(\tau + t) \in \mathbf{I}$ a.s., the a.s. left continuity of the sample function at time $\tau + t$ implies that the indicator function of $\{X(J_n + t) = j\}$ a.s. converges to that of $\{X(\tau + t) = j\}$. Therefore, from the lemma

$$P\{X(J_n + t) = m \mid \underline{\underline{H}}_{J_n}\} \to P\{X(\tau + t) = j \mid \underline{\underline{H}}_{\tau-}\} \qquad \text{a.s.}$$

The first member is equal to $p_{\chi_n(\cdot),j}(t)$, and the second one is equal to $\xi_j^a(t)$.

Using this result, we can find a subset A of Ω such that A^c is negligible, and, for every $\omega \in A$

$$p_{\chi_n(\omega),j}(t) \to \xi_j^a(t) \qquad \text{for all } j \in \mathbf{I} \text{ and rational } t > 0.$$

Remark that the first member is (for fixed ω) an entrance law. Proposition 2 will then follow from the following lemma.

LEMMA. *Let* $(g_j^n(t))$, $(g_j(t))$ *be normalized entrance laws. Assume that*

$$g_j^n(t) \underset{n\to\infty}{\to} g_j(t) \qquad \text{for all } j \text{ and rational } t > 0.$$

Then $g_j^n(\cdot) \to g_j(\cdot)$ *on* T^0 (and in fact uniformly on compact sets).

Proof: Let t be rational. We have for all $s > 0$

$$g_j^n(t+s) = \sum_k g_k^n(t) p_{kj}(s) \ .$$

Therefore, from Fatou's lemma

$$g_j(t+s) \leqq \liminf g_j^n(t+s) \ .$$

But

$$1 = \sum_j g_j(t+s) \leqq \sum_j \liminf g_j^n(t+s) \leqq \liminf \sum_j g_j^n(t+s) = 1 \ .$$

We have therefore for all $r > 0$ and j

$$g_j(r) = \liminf g_j^n(r) \ .$$

Now this implies the same result with lim instead of lim inf. Indeed, take any r, any j, and extract a subsequence n_i such that

$$\lim_i g_j^{n_i}(r) = \lim_n \sup g_j^n(r) \ .$$

Then the preceding result also applies to this subsequence, giving

$$g_j(r) = \lim_i \inf g_j^{n_i}(r) = \lim_n \sup g_j^n(r) \ .$$

The lemma is proved.

DEFINITIONS. 1) *Two boundary atoms* a *and* b *are indistinguishable iff* $\xi_j^a(t) = \xi_j^b(t)$ *for all* j, t.*

* A boundary atom a can also be "indistinguishable" from a state $i \in I$, in the sense that $\xi_j^a(t) = p_{ij}(t)$ for all j, t. However, in this case we *do* distinguish them.

2) *A (passable) boundary point is an equivalence class of indistinguishable boundary atoms. The finite set of all boundary points is denoted by* **B**, *and called the boundary.*

3) *If* a *is a boundary point,* a *is a finite set of indistinguishable boundary atoms* $a_1, \ldots, a_n$. *We then set*

$$\xi_j(t) = \xi_j^{a_i}(t) \qquad \textit{for all } j, t;$$

$$\Delta^a = \Delta^{a_1} \cup \cdots \cup \Delta^{a_n} \quad ;$$

$$\tau^a = \tau_{\Delta^a} = \tau^{a_1} \wedge \cdots \wedge \tau^{a_n} \quad .$$

The optional r.v.'s $\tau^a_{(t)}$ *are defined similarly. The probability law on* Ω *corresponding to the entrance law* $(\xi_j^a(t))$ *will be denoted by* P_a.

This terminology can be explained as follows: the state space I can be compactified in a natural way, so that all sample functions are right continuous and have left limits in the compactification I_c.* Then each boundary point a corresponds to exactly one point in the "boundary" $I_c \setminus I$ (which will be denoted also by a), the event Δ^a being a.s. identical with $\{\tau < \infty,\ X(\tau-) = a\}$ (the left limit being understood here in the sense of the topology of I_c). However, there may be several essentially different ways of of reaching a from the left, and these ways correspond to the "boundary atoms" or "exits to a": $a_1, \ldots, a_n$, belonging to a. We may draw a picture:

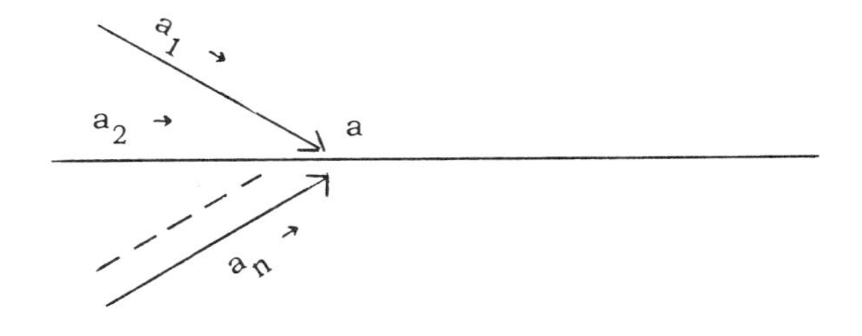

* See John Walsh, [16]. In general $I_c \setminus I$ is larger than the boundary **B** defined above!

This suggests also the following notations. For clarity, we may assume* (by throwing away a set negligible for all laws P) that $X(t) \neq \infty$ for every ω and every rational $t > 0$. Remember that $S_\infty(\omega)$ is a right closed set with empty interior. Let $L(\omega)$ be the set of all right endpoints of all I-intervals, that is, the set of all points of $S_\infty(\omega)$ which are left isolated. Then $L(\omega)$ is dense in $S_\infty(\omega)$ in the right topology. Let t belong to $L(\omega)$: then a non-empty I-interval lies to the left of t, and if we insert a rational s in this interval we have $t = \tau_{(s)}(\omega)$. Therefore

$$L(\omega) = \{\tau_{(s)}(\omega), \text{ s rational}\} .$$

Now the value $\tau_{(s)}(\omega)$ is the value of one and only one of the r.v.'s $\tau^a_{(s)}(\omega)$. We may therefore decompose $L(\omega)$ into disjoint smaller sets, as follows.

DEFINITION. *For every boundary* exit *or boundary* point a, *we set*

$$\tilde{S}_a(\omega) = \{\tau^a_{(s)}(\omega), \text{ s rational} > 0\}$$

and we denote by $S_a^+(\omega)$ *the set of all limit points from the right of* $\tilde{S}_a(\omega)$, *by* $S_a^-(\omega)$ *the set of all its limit points from the left.*

Of course, if a is a boundary point, and if $a_1, \ldots, a_n$ are the corresponding indistinguishable exits, then $\tilde{S}_a(\omega)$ is the union of $\tilde{S}_{a_1}(\omega) \ldots \tilde{S}_{a_n}(\omega)$.

NOTATION. $t \in \tilde{S}_a(\omega)$ *will be written* $X_{t-}(\omega) = a$, *whenever* a *is a boundary exit or a boundary point.*

PROPOSITION 3. *For any two distinct boundary points* a, b, *we have for a.e.* $\omega \in \Omega$ *(for every law* P*)*

$$S_a^+(\omega) \cap S_b^+(\omega) = \emptyset .$$

---

* This assumption, however, is made only for clarity here, and will not be made in the sequel. It isn't required for the definition below.

(However, it *may* happen that $\tilde{S}_a(\omega) \cap S_b^+(\omega) \neq \emptyset$. It can also be proved that $S_a^- \cap S_b^-(\omega) \neq \emptyset$ almost surely.)

Proof: Let Ω' be the set of all $\omega \in \Omega$ such that, for every A rational, every $j \in \mathbf{I}$, the martingale sample function

$$Y_{A,j}(s,\omega) = p(A-s, X_s(\omega), j) \qquad (s \in \,]0,A[)$$

has right and left limits along the rationals. The complement of Ω' has probability 0 for every law P. On the other hand, according to Proposition 2, for every t rational we have a.s.

$$Y_{A,j}(\tau^a_{(t)}-,\omega) = \xi^a_j(A-\tau^a_{(t)}(\omega)) \quad \text{on } \{\tau^a_{(t)} < A\} .$$

Deleting a new exceptional set of measure 0, we get a set Ω''_a of full measure such that, for $\omega \in \Omega''_a$

$$s \in \tilde{S}_a(\omega) \implies Y_{A,j}(s-,\omega) = \xi^a_j(A-s) \qquad \text{(every rational } A > s,\ j \in \mathbf{I}).$$

Assume $s \in S_a^+(\omega)$: then the right limit $Y_{A,j}(s+,\omega)$ is the right limit of the left limits, and since $\xi^a_j(\cdot)$ is continuous,

$$s \in S_a^+(\omega) \implies Y_{A,j}(s+,\omega) = \xi^a_j(A-s) \quad .$$

Now $\Omega''_a \cap \Omega''_b$ also is a set of full measure, and if ω belongs to this set,

$$s \in S_a^+(\omega) \cap S_b^+(\omega) \implies \xi^a_j(A-s) = \xi^b_j(A-s) \qquad \text{for all rational } A > s,\ j \in \mathbf{I}$$

which is impossible, since a and b are distinct boundary *points*, which means that the corresponding entrance laws are distinct.

DEFINITION. *Let* a *and* b *be two boundary exits (* a *may be equal to* b*) or two boundary points. Then we say that* a *sticks to* b *iff*

$$P_a\{\inf \tilde{S}_b = 0\} \neq 0.$$

We say that a *is sticky iff* $P_a\{\inf \tilde{S}_a = 0\} = 1$.

Obviously, a sticks to b if and only if for some measure μ, $P_\mu\{\forall\ \varepsilon > 0,\ ]\tau^a, \tau^a + \varepsilon[\ \cap \tilde{S}_b \neq \emptyset\} > 0$. We shall see later on that a is sticky $\Longleftrightarrow$ a sticks to a.

If a and b denote two boundary points, we say that a sticks to b iff some exit a' corresponding to a sticks to some exit b' corresponding to b. Then *every* a' corresponding to a sticks to *that* b' (since $P_{a'}$ is the same law for every a' corresponding to a), but not necessarily to *every* b' corresponding to b.

PROPOSITION 4. *If* a *is sticky, for a.e.* ω, $\tilde{S}_a(\omega)$ *is dense in itself (more precisely,* $\tilde{S}_a(\omega) \subset S_a^+(\omega)$*). If* a *is non-sticky, for a.e.* ω, $\tilde{S}_a(\omega)$ *is isolated (in particular,* a *cannot stick to* a*). If* a *sticks to* b, *and* b *is distinguishable from* a, *then* b *is sticky and* a *is nonsticky.*

Proof: If a is sticky, the value τ^a a.s. belongs to S_a^+. Applying the ordinary Markov property, we get that $\tau^a_{(t)}$ belongs a.s. to S_a^+ for every rational t, that is, $\tilde{S}_a(\omega) \subset S_a^+(\omega)$.

The second statement is more difficult to prove. We shall make computations with the law P_a: since it is the normalized law of the post τ^a-process for every law P, our result will be valid for arbitrary P.

Let us choose $\varepsilon > 0$ such that

$$P_a\{]0, \varepsilon[\ \cap \tilde{S}_a \neq \emptyset\} < 1$$

and denote by λ this probability. Let r be a rational number > 0; let us set

$$U_1 = r + \tau^a \circ \theta_r \quad (\text{"first hit of a after r"}),\quad V_1 = U_1 + r$$

and note that $X(V_1) \in I$ a.s. on $\{V_1 < \infty\}$. Then

$$U_2 = V_1 + \tau^a \circ \theta_{V_1}, \quad V_2 = U_2 + r$$

and so on, inductively. Assume there are n points $t_1 < t_2 < \cdots < t_n$ in

$\tilde{S}_a(\omega) \cap]0,\varepsilon[$: then, if $r < t_1 \wedge \cdots \wedge (t_n - t_{n-1})$, we have $U_1 \leqq t_1, \ldots,$ $U_n \leqq t_n$. Therefore

$$P_a\{\text{there are at least } n \text{ points in }]0,\varepsilon[\cap \tilde{S}_a\} \leqq \lim_{r \to 0} P_a\{U_n < \varepsilon\} .$$

Now $P_a\{U_n < \varepsilon\} \leqq P_a\{U_1 < \varepsilon, U_2 - U_1 < \varepsilon \cdots U_n - U_{n-1} < \varepsilon\} \leqq \lambda^n$, which tends to 0 as $n \to \infty$. Therefore

$$P_a\{\text{the number of points in }]0,\varepsilon[\cap \tilde{S}_a \text{ is finite}\} = 1 .$$

The same result extends obviously to any law P_μ from the strong Markov property. Applying it now to the intervals $]0,\varepsilon[$, $[\varepsilon, 2\varepsilon[$, ... we find that

$$P_a\{\tilde{S}_a \text{ is discrete}\} = 1$$

and this is the second statement.

Assume now that a sticks to b. This implies that, for some measure μ, $P_\mu\{\tau^a \in S_b^+\} > 0$, and therefore $P_\mu\{\tilde{S}_a \cap S_b^+ \neq \emptyset\} > 0$. On the other hand, since a and b are distinguishable, $P_\mu\{S_a^+ \cap S_b^+ \neq \emptyset\} = 0$ (Proposition 3). Therefore $\tilde{S}_a$ isn't contained in S_a^+ (which implies that a is nonsticky), and $S_b^+ \neq \emptyset$ (which implies that b is sticky).

The Strong Markov Property at the Boundary

PROPOSITION 5. *Let* a *be a boundary point, and* T *be an optional r.v.*

1) *Assume that* a *is sticky, and that for* P*-a.e.* $\omega \in \{T < \infty\}$ *we have* $T(\omega) \in S_a^+(\omega)$. *Then*

$$P\{X(T+t) = j,\ T < \infty\} = P\{T < \infty\} \cdot \xi_j^a(t) . \tag{1}$$

2) *Assume that* a *is nonsticky, that* T *is predictable, and that for* P*-a.e.* $\omega \in \{T < \infty\}$ *we have* $T(\omega) \in \tilde{S}_a(\omega)$. *Then* (1) *holds true.*

Proof: We first recall some facts about predictable r.v.'s. This property *is relative to one given* law P: an optional r.v. T is predictable iff there exists an increasing sequence (S_n) of optional r.v.'s, a.s. converging to T, such that for every n, $S_n < T$ a.s. on $\{T > 0\}$. We then denote by $\underline{\underline{H}}_{T-}$ the σ-field $\vee_n \underline{\underline{H}}_{S_n}$ generated by the σ-fields $\underline{\underline{H}}_{S_n}$. It can be shown that it doesn't depend on the sequence (S_n) satisfying the above property. The following results are useful (see for instance "Guide détaillé" cited on p. 37).

1) If T is predictable, and $A \in \underline{\underline{H}}_{T-}$, T_A is predictable ($T_A = T$ on A $+\infty$ on A^c).

2) If $T_1, \ldots, T_n$ are predictable, $T_1 \wedge T_2 \wedge \cdots \wedge T_n$ is repdictable.

3) If (T_n) is an increasing sequence of predictable optional times, then $\lim T_n$ is predictable.

Now consider any $r > 0$ and define:

$$\begin{array}{ll} \text{if a is sticky, } U_0 = T + r, & V_1 = T + r + \tau^a \circ \theta_{T+r} ; \\ \text{if a is nonsticky, } U_0 = r, & V_1 = r + \tau^a \circ \theta_r , \\ U_1 = V_1 + r , & V_2 = U_1 + \tau^a \circ \theta_{U_1} , \\ U_2 = V_2 + r , & \ldots \text{ etc.} \end{array}$$

We have $X(U_i) \in I$ a.s. for all i, on $\{U_i < \infty\}$ (strong Markov property). Let us show that V_i is predictable. Using the strong Markov property at time U_i, we are reduced to showing that, for every law P_μ, τ^a is predictable: this follows from property 1) above, since $\tau^a = \tau_{\Delta^a}$, being predictable ($\tau = \lim J_n$), and Δ^a belonging to $\underline{\underline{H}}_{\tau-}$.

Using now the stronger statement of Proposition 1, we get that if

$A \in \underline{\underline{H}}_{V_i-}$ then

$$(2) \qquad P\{A, V_i < \infty, X(V_i + t) = j\} = P\{A, V_i < \infty\} \cdot \xi^a_j(t) .$$

(Proposition 1, however, is relative to a boundary *exit* a, not to a boundany *point* a: (2) is obtained by adding the equalities relative to all exits corresponding to a).

Now define T_r as follows:

– if a is sticky, and T is just optional, $T_r = V_1$;

– if a is nonsticky, and T is predictable,

$$T_r = \inf\{V_i(\omega) : V_i(\omega) \geqq T(\omega)\} .$$

In the latter case, we can write $T_r = \Sigma_i I_{\{V_i = T_r\}} V_i$; now $\{V_i = T_r\}$ belongs to $\underline{\underline{H}}_{V_i -}$ in both cases (easy proof left to the reader), and therefore we get from (2)

$$(3) \qquad P\{T_r < \infty, X(T_r + t) = j\} = P\{T_r < \infty\} \cdot \xi_j^a(t) .$$

So we must only prove that T_r tends to T from the right as $r \to 0$ (we do not assert that T_r decreases to T !). This follows at once from Proposition 3 if a is sticky, while $T_r = T$ for r small enough if a is nonsticky.*

REMARK. In the first case, let A denote an element of $\underline{\underline{H}}_T$; in the second case, an element of $\underline{\underline{H}}_{T-}$: then the theorem applies to T_A, giving

$$(4) \qquad P\{A, T < \infty, X(T + t) = j\} = P\{A, T < \infty\} \cdot \xi_j^a(t) .$$

That is, the usual form of the strong Markov property with conditional probabilities–the only nonstandard feature being the presence of $\underline{\underline{H}}_{T-}$ instead of $\underline{\underline{H}}_T$ in the nonsticky case.

The following proposition is an easy strengthening of the preceding one:

PROPOSITION 6. T *having the same meaning as in Proposition 5, denote by* S *a random variable* $\geqq$ T, *measurable with respect to* $\underline{\underline{H}}_T$ *in the first case, to* $\underline{\underline{H}}_{T-}$ *in the second case. Then*

$$(1) \qquad P\{X_S = j, S < \infty\} = E\{\xi_j^a(S - T), S < \infty\} .$$

---

* The reader should draw some pictures for himself.

Proof: We may write $S = T + U$, where U is positive, measurable with respect to $\underline{\underline{H}}_T$ ($\underline{\underline{H}}_{T-}$ in the second case). Then (1) is trivial from Proposition 5 and the preceding remark if U is countably valued, and we get the general case through a decreasing passage to the limit.

COROLLARY. *Set* $F(s) = P\{T \leqq s\}$. *Then for every* $t > 0$

$$P\{T < t, X(t) = j\} = \int_0^{t-} \xi_j^a(t-s)dF(s) \tag{2}$$

$$\text{(same result with } T \leqq t, \int_0^{t+}) .$$

Proof: Apply Proposition 6 with $S = t$ if $T < t$, $S = \infty$ if $T \geqq t$. Then we get from (1) that

$$P\{T < t, X(t) = j\} = E\{\xi_j^a(t-T), T < t\}$$

and this is just the right side of (2).

The Elementary Decomposition Formula

We shall give an easy application of the preceding corollary (in fact, it is far from requiring the full strength of the strong Markov property we have proved!). We first require some definitions.

DEFINITION. *For any boundary exit* a, $i \in \mathbf{I}$, *we set*

$$L_i^a(t) = P_i\{\tau^a \leqq t\} .$$

The row vector $L_{\cdot}^a(t)$ is denoted by $L^a(t)$, and we obviously have from probabilistic interpretation

$$L^a(s+t) = L^a(t) + \Phi(t)L^a(s) .$$

Therefore, from Proposition 2 of Chapter I, we can write

$$L_i^a(t) = \int_0^t \ell_i^a(s)\,ds$$

where $(\ell_i^a(s))$ *is an exit law for the minimal semi-group.**

PROPOSITION 7. *We have*

$$(1) \qquad p_{ij}(t) = f_{ij}(t) + \sum_{a \in E} \int_0^t L_i^a(ds)\,\xi_j^a(t-s)$$

$$= f_{ij}(t) + \sum_{a \in E} \int_0^t \ell_i^a(s)\,ds\,\xi_j^a(t-s) \; .$$

Proof: We express that $P_i\{X(t) = j\}$ is the union of the disjoint events $P_i\{X(t) = j,\ \tau > t\}$ and $P_i\{X(t) = j,\ \tau \leqq t,\ \tau = \tau^a\}$, and we apply to the predictable optional time $\tau^a = \tau_{\{\tau=\tau^a\}}$ the corollary to Proposition 6. Note that finiteness of the boundary was not used in this proof, therefore (1) holds for a countable E.

We shall now write a more concise version of formula (1). We introduce the row vector $\xi^a(t)$, the column vector $\ell^a(t)$, and the Laplace transforms

$$\hat{\Pi}(\lambda) = \int_0^\infty e^{-\lambda t}\,\Pi(t)\,dt\,, \qquad \hat{\Phi}(\lambda) = \int_0^\infty e^{-\lambda t}\,\phi(t)\,dt\,,$$

$$\hat{\ell}^a(\lambda) = \int_0^\infty e^{-\lambda t}\,\ell^a(t)\,dt\,, \qquad \hat{\xi}(\lambda) = \int_0^\infty e^{-\lambda t}\,\xi^a(t)\,dt\;.$$

More generally, $\hat{}$ will be used to denote Laplace transforms. Then (1) can be written as

$$(2) \qquad \hat{p}_{ij}(\lambda) = \hat{f}_{ij}(\lambda) + \sum_a \hat{\ell}_i^a(\lambda)\,\xi_j^a(\lambda)\;.$$

---

* It can be shown that these exit laws generate extremal rays of the cone of exit laws for the minimal semi-group.

In standard matrix notation

$$\hat{\Pi}(\lambda) = \hat{\Phi}(\lambda) + \sum_a \hat{\ell}^a(\lambda) \otimes \hat{\xi}^a(\lambda) \quad . \tag{3}$$

Some comments are now in order to understand what is going on. Our main purpose consists in describing all the STM's which admit the coefficients q_i and q_{ij}—otherwise stated, all the STM's with the same minimal semi-group Φ. To achieve this, we must find a formula giving $\Pi(t)$ as a function of the minimal semi-group Φ and some "parameters" which may be numbers, measures, or entrance laws and exit laws *with respect to the minimal semi-group*. Formula (1) and (3)) above is a first step in that direction, because it gives us a closed formula allowing the computation of Π using Φ (the minimal semi-group), the set of exits (which can be deduced intrinsically from the knowledge of the minimal semi-group; incidently, this is the reason why we are dealing with boundary *exits* instead of boundary *points*: indistinguishability is a property which refers to Π, not to Φ), and the Φ-exit laws $\ell^a(t)$. On the other hand, it also uses the entrance laws $\xi^a(t)$, and these are relative to Π, not Φ, Therefore formula (1) isn't yet satisfactory, and reduction must be carried further. This will be the purpose of the next two sections.

Section 2. The First Switching Time

We are going now to single out one boundary point a, and study closely the behavior of the process "at a". The natural way to do it consists in killing the process at the first time it hits a boundary point different from

DEFINITIONS. 1) $\beta^a(\omega) = \inf\{t > 0 : t \in \bigcup_{b \in B \setminus \{a\}} \tilde{S}_b(\omega)\}$.

This is an optional r.v., and even a *terminal time*: if $t < \beta^a$, then $\beta^a = t + \beta^a \circ \theta_t$.

If we start at a (i.e., if the probability law is P_a), then β^a can be interpreted as the first time where a *change of boundary occurs.*

2) Killing the process at β^a, we get a new matrix function $\Pi^a(t)$ with coefficients

$$p^a_{ij}(t) = P_i\{X(t) = j, t < \beta^a\} .$$

The fact that β^a is a terminal time implies (by reasonings quite standard in the theory of Markov processes) that $\Pi^a(\cdot)$ is a SSTM, and the probabilistic interpretation implies obviously that it has the same coefficients q_i, q_{ij} as Π and $\emptyset$. We obviously have $\Phi \leqq \Pi^a \leqq \Pi$.

3) If $F(\cdot)$ is any entrance law relative to Π, we get an entrance law $G(\cdot)$ relative to Π^a by setting

$$g_j(t) = P_F\{X(t) = j, t < \beta^a\}$$

where P_F is the law on Ω corresponding to F. Taking for F the entrance law $(\xi^a_j(t))$, we get the following Π^a- entrance law, which is very important:

$$\rho^a_j(t) = P_a\{X(t) = j, t < \beta^a\} .$$

For a given a, it is possible that $\rho^a_j(t) = 0$ for every j and t; such a boundary point is called *ephemeral.*

4) We are going now to define the distribution of the switching time to a given boundary point. This requires some care, because the sample function, after some time spent without hitting any boundary except a, can switch to some nonsticky boundary c, and then (if c sticks to b) switch again instantaneously to b. This can happen only if b is sticky (Proposition 4). Of course the "true" switching value in this case is c, not b.

We define $F^{aa}(t) = 0$ for all t.

If $b \neq a$ is nonsticky, we define $F^{ab}(t) = P_a\{\beta^a \leqq t, \beta^a \in \tilde{S}_b\}$.

If $b \neq a$ is sticky, we define

$$F^{ab}(t) = P_a\{\beta^a \leqq t, \beta^a \notin \tilde{S}_c \text{ for every nonsticky } c \neq a, \beta^a \in S^+_b\}$$

We can have $F^{ab}(0) > 0$ only if a is nonsticky and b is sticky. In that case, the measure $dF^{ab}(\cdot)$ has a mass at 0 equal to $F^{ab}(0)$. Except for this mass, the measure is continuous, since the mass at $t > 0$ is smaller than $P_a\{X(t) \notin \mathbf{I}\} = 0$. Accordingly, we need not specify if the bounds are included in integrals with respect to this measure.

It is understood that an integral such as $\int_0^t$ *is over* $[0, t]$.

PROPOSITION 8. *We have*

$$\sum_{i \in \mathbf{I}} \rho_i^a(t) + \sum_{b \in \mathbf{B}} F^{ab}(t) = 1 \quad ; \tag{1}$$

$$\rho_i^a(t) + \sum_{b \in \mathbf{B}} \int_0^t F^{ab}(ds)\xi_i^b(t-s) = \xi_i^a(t) \quad : \tag{2}$$

Proof: Formula (1) is obvious from the probabilistic interpretation. For any $b \neq a$, denote by C_b the optional r.v. $(\beta^a)_{A_b}$, where A_b is the event (belonging to $\underline{\underline{H}}_{\beta^a}$) defined as follows:

if b is nonsticky, $A_b = \{\beta^a \in \tilde{S}_b\}$

if b is sticky, $A_b = \{\beta^a \notin \tilde{S}_c$ for every nonsticky $c \neq a$, $\beta^a \in S_b^+\}$.

Now, according to Propositions 5-6 and their corollary, we need only prove that C_b is predictable for every nonsticky b (relative to the law P_a). We shall use the following lemma:

LEMMA. *Let* C *be an optional r.v.; assume there exists a sequence of predictable optional times* C_n *such that* $\{C < \infty\} = \cup_n \{C = C_n < \infty\}$ *and* $\{C = C_n\} \in \underline{\underline{H}}_{C_n-}$ *for all* n. *Then* C *is predictable.*

(Proof: The set of all $A \in \underline{\underline{H}}_C$ such that C_A is predictable is closed under countable unions (see for instance "Guide détaillé"). Set $A_n = \{C = C_n\}$: $C_{A_n} = (C_n)_{A_n}$ * is predictable since C_n is predictable and

* Recall that $C_{A_n} = \infty$ on A_n^c and $= C$ on A_n.

$A_n \in \underline{\underline{H}}_{C_n-}$. Therefore, if we set $A = \cup A_n$, C_A is predictable. Finally, $C = C_A$ is predictable.)

Let us apply the lemma, with $C = C_b$ (b nonsticky) the optional times C_n being replaced by the times $\tau^b_{(r)}$, r rational. We need only show that $\{C_b = \tau^b_{(r)}\} \in \underline{\underline{H}}_{\tau^b_{(r)}-}$. This event is just $\{\beta^a = \tau^b_{(r)}\}$, and we cannot have $\beta^a > \tau^b_{(r)}$ from the definition of β^a as the first time some boundary point other than a is hit: therefore $\{\beta^a = \tau^b_{(r)}\}$ is the complement of $\{\beta^a < \tau^b_{(r)}\}$, which belongs to $\underline{\underline{H}}_{\tau^b_{(r)}-}$. Proposition 8 is proved.

Some Algebra

We are returning now to the situation of the end of Section 1. We have seen there that we could "compute" Π if we knew the entrance laws $\xi^a_j(t)$ for all boundary exits. This amounts to knowing which boundary exits are indistinguishable, and knowing the entrance laws corresponding to the boundary points. We shall see now that we can "compute" these entrance laws if we know for all boundary points 1) the measures $F^{ab}(dt)$, 2) the entrance laws $\rho^a_j(t)$. Since these are entrance laws with respect to Π^a not Π, we shall be one step closer to the complete decomposition of Π relative to $\emptyset$.

We shall use some Laplace transform notations, as follows:

$$\hat{F}^{ab}(\lambda) = \int_0^\infty e^{-\lambda t} F^{ab}(dt) \quad , \qquad (\lambda > 0)$$

$$\hat{\rho}^a_j(\lambda) = \int_0^\infty e^{-\lambda t} \rho^a_j(t)dt \quad .$$

But this time we shall consider $\hat{F}$ as a $\mathbf{B} \times \mathbf{B}$ matrix, and denote by $\hat{\xi}_j$, $\hat{\rho}_j$ the columns vectors $(\hat{\xi}^a_j)_{a \in \mathbf{B}}$, $(\hat{\rho}^a_j)_{a \in \mathbf{B}}$—that is, we are doing matrix computations with $\mathbf{B}$, not $\mathbf{I}$ as usual, as index set.

PROPOSITION. 9. *We have*

$$(1) \qquad (I - \hat{F}(\lambda))\hat{\xi}_j(\lambda) = \hat{\rho}_j(\lambda)$$

and $I - \hat{F}(\lambda)$ *is invertible.*

Proof: Formula (1) is obtained for every $\lambda > 0$ by taking the Laplace transform of formula (2) in Proposition 8. To prove that $I - \hat{F}(\lambda)$ is invertible, we shall use an algebraic lemma (cf. O. Taussky, "A recurring theorem on determinants," *American Mathematical Monthly*, Vol. 56 (1949), pp. 672-676).

LEMMA. *Let* A *be a substochastic matrix on a finite set* B. *Then* $I - A$ *is noninvertible if and only if there exists a subset* C *of* B *such that* $A|_{C \times C}$ (A restricted to $C \times C$) *is stochastic.*

Let us assume that there is $C \subset B$ such that $\hat{F}(\lambda)|_{C \times C}$ is stochastic, and derive a contradiction from it. First of all, since $\hat{F}(\lambda)$ has zeroes on the diagonal, C cannot have just one element. On the other hand, let a be any element of C. We then have $\Sigma_{b \in C} \hat{F}^{ab}(\lambda) = 1$: since this is the value at λ of the Laplace transform of the measure $\Sigma_{b \in C} dF^{ab}(t)$, which has total mass $\leqq 1$, equality to 1 is possible only if this measure is a unit mass at 0, and therefore

there exists for every $a \in C$ some $b \in C$ such that $F^{ab}(0) > 0$.

Now this implies that a is nonsticky and b is sticky (Proposition 4). Therefore every $a \in C$ is nonsticky and some $b \in C$ is sticky, the desired contradiction.

To complete the program described in the last lines of Section 1, we need only find a description of the Π^a entrance law ρ^a, relative to the minimal semi-group. This will be done in Section 3.

Section 3. The Main Decomposition

We are going to recall some definitions concerning recurrence of ordinary states in I. We use notations relative to Π, but the results will often be applied to Π^a or Φ.

NOTATION. Let i and j belong to I. Then $i \rightsquigarrow j$ means that $p_{ij}(t) > 0$ for some $t > 0$ (that j can be reached from i); $i \leftrightsquigarrow j$ means that $i \rightsquigarrow j$ and $j \rightsquigarrow i$.

Define successive optional r.v.'s as follows: $T_0 = 0$, $S_0 = \inf\{t : X(t) \neq i\}$, $T_1 = \inf\{t > S_0 : X(t) = i\}$, $S_1 = \inf\{t \geqq T_1 : X(t) \neq i\}$ and so on. We deduce easily from the strong Markov property that there exist positive, independent, identically distributed variables $V_1, \ldots, V_n$ (possibly $+\infty$) such that, if Ω is given the law P_i, T_n is identical in law to $V_1 + \cdots + V_n$. According to standard renewal theory:

- either $P_i\{V_i = \infty\} > 0$, and then a.s. $T_n = \infty$ for n sufficiently large,
- or $P_i\{V_i = \infty\} = 0$, and then a.s. T_n is finite for all $n \geqq 0$ and tends to ∞ with n.

In the first case, i is called a nonrecurrent state, and in the second case, a recurrent state. Otherwise stated:

DEFINITION. i is recurrent iff $P_i\{S_i$ is unbounded$\} = 1$; i is nonrecurrent iff $P_i\{S_i$ is unbounded$\} = 0$.

Denote by H_i the time spent in state i between T_n and T_{n+1} (or between T_n and S_{n+1}). These random variables are strictly positive with probability one on $\{T_n < \infty\}$, independent, identically distributed (exponential distribution with parameter q_i). In the recurrent case, the total time spent at i, equal to ΣH_n, is therefore a.s. infinite. On the other hand, in the nonrecurrent case, the total time spent at i is equal to $H = \Sigma_{n=0}^{N-1} H_n$ where N is the first integer n such that $T_n = \infty$. It is well known that N is independent of the H_n's, and that $N-1$ has a Pascal (geometric) distribution. Morevoer, $E[H] = E[H_1]E[N-1] < \infty$. Therefore, we have the following classical criteria for recurrence.

RECURRENCE CRITERION. *If* i *is recurrent:*

- P_i-a.s. $S_i(\omega)$ *is unbounded,*
- P_i-a.s. $m(S_i(\omega)) = \infty$ (m = Lebesgue measure on the line),
- $\int_0^\infty p_{ii}(t)\,dt = +\infty$ (this is equal to $E_i[m(S_i)]$).

If i *is nonrecurrent:*

- P_i-a.s. $S_i(\omega)$ *is bounded,*
- P_i-a.s. $m(S_i(\omega)) < \infty$,
- $\int_0^\infty p_{ii}(t)\,dt < \infty$.

We shall need one more result, which will not be proved here (see [1]).

PROPOSITION. *If* i *is recurrent, and* $i \rightsquigarrow j$, *then* j *is recurrent and* $j \rightsquigarrow i$. *Let* R *be the set of all recurrent states; then (a.s.) every sample function remains in* R *after the first time it hits* R.

In the remainder of this chapter, we shall denote by $\mathbf{I}_r$, $\mathbf{I}_t$ the sets of recurrent, nonrecurrent states *relative to* $\emptyset$. From our point of view, $\mathbf{I}_r$ is uninteresting: if $i \in \mathbf{I}_r$, τ is P_i-a.s. infinite (since τ is the lifetime of the minimal process).

We are going first to define the symbol $\rightsquigarrow$ and recurrence for boundary points.

DEFINITION. If $a \in \mathbf{B}$, $i \in \mathbf{I}$, $a \rightsquigarrow i$ means that $\xi_i^a(t) > 0$ for some $t > 0$. If $a \in \mathbf{B}$, $i \in \mathbf{I}$, $i \rightsquigarrow a$ means that $P_i\{\tilde{S}_a \neq \emptyset\} > 0$. If $a \in \mathbf{B}$, $b \in \mathbf{B}$, $a \rightsquigarrow b$ means that $F^{ab}(t) > 0$ for some t.

It is easily seen that the relation is transitive. If C is a set, $a \rightsquigarrow C$ means that $a \rightsquigarrow$ some $c \in C$. The negative of $\rightsquigarrow$ is $\not\rightsquigarrow$.

DEFINITION. $a \in \mathbf{B}$ is *recurrent* iff $P_a\{\tilde{S}_a$ is unbounded$\} = 1$.
$a \in \mathbf{B}$ is a *trap* iff $a \not\rightsquigarrow \mathbf{B} \setminus \{a\}$.

Lemmas Concerning Recurrence

LEMMA 1. *If* $j \in \mathbf{I}_t$ *and* $j \not\rightsquigarrow \mathbf{B}$, *then* j *is nonrecurrent for* Π.

Indeed, since $j \not\rightsquigarrow \mathbf{B}$, we have

$$\int_0^\infty p_{jj}(t)\,dt = \int_0^\infty f_{jj}(t)\,dt < \infty\ .$$

LEMMA 2. *If* $j \in \mathbf{I}$ *is* Π-*recurrent and* $j \rightsquigarrow a \in \mathbf{B}$, *then* a *is recurrent.*

Let us choose $s > 0$ such that $P_j\{\tilde{S}_a \cap [0, s[\neq \emptyset\} = h > 0$ and define

$$R_0 = 0,\ R_1 = \inf\{t > s :\ X(t) = j\}, \ldots, R_{n+1} = \inf\{t > R_n + s :\ X(t) = j\}\ .$$

Since j is recurrent, all R_n are finite a.s.-P_j and $R_n \to \infty$ a.s.-P_j. We also have $P_j\{\tilde{S}_a \cap [R_n, R_{n+1}[\neq \emptyset\} \geqq P_j\{\tilde{S}_a \cap [0, s[\neq \emptyset\} = h > 0$. The events in the first member are all independent and have the same probability: according to Borel-Cantelli

$$P_j\{\tilde{S}_a \cap [R_n, R_{n+1}[\neq \emptyset \text{ for infinitely many } n\} = 1\ . \tag{1}$$

Applying the strong Markov property at time τ^a, the same result holds for P^a instead of P_j and therefore a is recurrent.

LEMMA 3. *If* $j \in \mathbf{I}$ *is nonrecurrent for* Π, *we have* $\int_0^\infty \xi_j^a(t)\,dt < \infty$ *for all* $a \in \mathbf{B}$.

We know that $\int_0^\infty p_{jj}(t)\,dt < \infty$; denote this number by M. Then the maximum principle implies that $\int_0^\infty p_{ij}(t)\,dt \leqq M$. Let $\mu_i(t)$ be any normalized entrance law. Integrating with respect to $\mu_i(\varepsilon)$ we get

$$\int_0^\infty \sum_i \mu_i(\varepsilon)\,p_{ij}(t)\,dt \leqq M, \quad \text{or} \quad \int_\varepsilon^\infty \mu_j(t)\,dt \leqq M.$$

Finally, we let $\varepsilon \to 0$.

The next lemma is crucial.

LEMMA 4. *If* $j \in I$, $j \rightsquigarrow B\setminus\{a\}$ $(a \in B)$, *then* $\int_0^\infty \rho_j^a(t)\,dt < \infty$.

Let us choose $b \in B\setminus\{a\}$, and $s > 0$ such that $P_j\{\tilde{S}_b \cap [0, s[\, \neq \emptyset\} = h > 0$. Then $\rho_j^a(ns)h$ is the probability, starting at a, of being at j at time ns without having hit any other boundary than a, and of hitting b between ns and $(n+1)s$. Therefore, these are probabilities of *disjoint* events, and we have

$$\sum_{n=1}^{\infty} \rho_j^a(ns)h \leqq 1 \qquad \text{or} \qquad \sum_{n=1}^{\infty} \rho_j^a(ns) \leqq \frac{1}{h} \quad .$$

This is a Riemann sum approximating $\int_0^\infty \rho_j^a(t)\,dt$. To majorize precisely the integral, we notice that $\rho_j^a(t) f_{jj}(u) \leqq \rho_j^a(t+u)$. Therefore, if t belongs to $[ks, (k+1)s[$ we have

$$\rho_j^a(t) \leqq M\rho_j^a((k+1)s) \qquad \text{with} \qquad M = 1/\inf_{0 \leqq v \leqq s} f_{jj}(v) \ ,$$

and the lemma follows at once.

LEMMA 5. *If* $a \in B$ *is recurrent, and* $a \rightsquigarrow j \in I$, *then* j *is recurrent and* $j \rightsquigarrow a$. *Let* I^a *be the set of all* $j \in I$ *such that* $a \rightsquigarrow j$, *then* I^a *is a recurrent class: all states in* I^a *are recurrent and communicate. If* $j \in I^a$ *is such that* $j \rightsquigarrow k \in I$, *then* $k \in I^a$. *We have* $\int_0^\infty \xi_j^a(t)\,dt = \infty$ *for all* $j \in I^a$.

Proof: 1) *If* a *is recurrent and* $j \not\rightsquigarrow a$, *then* $a \not\rightsquigarrow j$.

Since $j \not\rightsquigarrow a$, $X_t = j$ implies that a is not hit after t. Therefore

$$\xi_j^a(t) \leqq P_a\{\tilde{S}_a \cap\,]t, \infty[\, = \emptyset\} \ .$$

This probability is 0 since a is recurrent, and 1) is proved.

2) *All states in* I^a *communicate.*

Take j and k in I^a. Then $a \rightsquigarrow j$, $a \rightsquigarrow k$; from 1), $j \rightsquigarrow a$, $a \rightsquigarrow k$, therefore $j \rightsquigarrow k$, and $k \rightsquigarrow j$ by symmetry.

3) $j \in I^a$, $j \rightsquigarrow k \Longrightarrow k \in I^a$ (transitivity).

4) *Every* $j \in I^a$ *is recurrent* (with respect to Π).

This is a little more difficult. Choose $s > 0$ such that $\xi^a_j(s) = h > 0$ and set $R_0 = 0$, $R_1 = \inf\{t > 2s : t \in \tilde{S}_a\}, \ldots, R_{n+1} = \inf\{t > R_n + 2s : t \in \tilde{S}_a\}$. Since a is recurrent, R_n is finite and increases to ∞ with n. We have from the strong Markov property, since $X(R_n + s) = j$ implies the event in the first member below,

$$P_a\{S_j \cap]R_n, R_{n+1}[\neq \emptyset \mid \underline{\underline{H}}_{R_n-}\} \geqq \xi^a_j(s) \ .$$

Using now Lévy's form of the Borel-Cantelli Lemma (Chapter II, lemma after Proposition 12: the event $\{S_j \cap]R_n, R_{n+1}[\neq \emptyset\}$ belongs to $\underline{\underline{H}}_{R_{n+1}-}$), we get

$$P_a\{S_j \cap]R_n, R_{n+1}[\neq \emptyset \text{ for infinitely many } n\} = 1 \ ,$$

and this implies by 1) the same property for P_j : j is recurrent.

5) *If* $j \in I$ *is recurrent and* $j \rightsquigarrow a$, *then* $a \rightsquigarrow j$ (this does not require the hypothesis that a is recurrent, in fact, it implies that a is recurrent by Lemma 2). Indeed, let us assume that $a \not\rightsquigarrow j$, then j cannot be hit after τ^a: since j is recurrent, $P_j\{\tau^a < \infty\} = 0$, and $j \not\rightsquigarrow a$.

6) $$\int_0^\infty \xi^a_j(t)\,dt = \infty \textit{ for all } j \in I^a \ (0 \textit{ for } j \notin I^a!) \ .$$

Choose some $s > 0$ such that $\xi^a_j(s) > 0$, and define

$$R_0 = 0, \ldots, R_{n+1} = \inf\{t > R_n + s : t \in \tilde{S}_a\} \ .$$

Then

$$E_a[m(S_j \cap [R_n, R_{n+1}[)] = E_a\left[\int_{R_n}^{R_{n+1}} I_{\{j\}} \circ X_u\, du\right] \geqq \int_0^s \xi^a_j(t)\,dt > 0$$

(since $\xi^a_j(\cdot)$ is continuous, and strictly positive at s). The random variables $m(S_j \cap [R_n, R_{n+1}[)$ are independent and identically distributed, and strictly positive with probability > 0 (measure P_a). According to the Borel-Cantelli lemma, their sum is a.s. infinite, and this sum is $m(S_j)$. We have thus obtained a result stronger than 6).

The Main Theorem: Statement

Recall (end of Section 2) that the construction of Π starting from Φ is now reduced to the construction of the entrance law ρ^a for each boundary point a. This is done in the following statement, which will be later on split into smaller propositions for the purpose of the proof.

THEOREM 1. *To each boundary point* a *which is not ephemeral we can associate a positive measure* E^a *on* T, *and a (non-normalized*) entrance law* $(\eta_j^a(t))$ *relative to* Φ, *such that*

$$(1) \qquad \rho_j^a(t) = \int_0^t E^a(ds)\eta_j^a(t-s)^{**} \qquad \textit{for all } t > 0 .$$

Moreover

1) E^a *is bounded if and only if* a *isn't a recurrent trap;*

2) E^a *is absolutely continuous except possibly for an atom at* 0 (we shall see later on that such an atom occurs if and only if a is nonsticky).

The first step in the proof is the following proposition, which is a weakened form of (1).

PROPOSITION 10. *With the same notations as above, we can find* E^a *and* $(\eta_j^a(t))$ *such that* (1) *holds for almost every (Lebesgue measure)* $t \in T$. E^a *is bounded if* a *isn't a recurrent trap.*

---

* We shall see later on (Proposition 6 of next chapter) that $<\eta^a(t),1> < \infty$ for $t > 0$ (this is easy), but $<\eta^a(t),1> \underset{t\to 0}{\to} \infty$ if a is sticky.

** Recall that $\eta_j^a(0+)$ exists (Chapter I). In any case, it doesn't matter whether we integrate over $[0,t[$ or $[0,t]$.

The proof will require an analytical lemma.

LEMMA 6. *Let* σ *be a function on* T, *positive, nonidentically* 0, *finite for all* $t > 0$, *decreasing right continuous, such that* $\sigma(t) \to 0$ *as* $t \to \infty$ and $\int_0^1 \sigma(t)\,dt < \infty$. *Let* c *be a number* $\geqq 0$. *There exists a positive measure (unique)* $E(ds)$ *on* T *such that for a.e.* t *

$$\int_0^t [c + \sigma(t-s)]E(ds) = 1 \quad . \tag{1}$$

Moreover

1) The total mass of E is $1/c$.

2) Assume $\sigma(0+) < \infty$. Then (1) holds everywhere and E has an atom at 0, equal to $1/(c + \sigma(0+))$.

3) Assume $d\sigma$ is absolutely continuous. Then (1) holds for all $t > 0$, and E is absolutely continuous, except for the atom at 0 mentioned in 2) if $\sigma(0+) < \infty$.

The proof will not be given here (it will be given in the Appendix). However, it is interesting to note here that multiplying (1) by $e^{-\lambda t}$ and integrating, we get that the Laplace transform $\hat{E}$ of E satisfies

$$\hat{E}(\lambda)\left[\frac{c}{\lambda} + \hat{\sigma}(\lambda)\right] = \frac{1}{\lambda} \quad . \tag{2}$$

Therefore, $\hat{E}$ is finite and known (uniqueness of E follows at once). This is of course the easiest path to existence also.

Proof of Proposition 10 in the Not Recurrent Trap Case:

We shall need two methods to prove Theorem 1, according to whether

---

* In fact, for all $t > 0$, as proved very recently by H. Kesten in a long paper, Memoirs of, *Amer. Math. Soc.*, No 93. A purely analytic and shorter proof has since been given by L. Carleson (private communication).

a is a recurrent trap or not. We shall first assume that a isn't a recurrent trap—i.e., that *either* a isn't recurrent, *or* if recurrent it isn't a trap. We shall also introduce the following:

AUXILIARY HYPOTHESIS. I_r *is empty: there are no* Φ*-recurrent states.*

We shall see, after the proof of Proposition 10 (in the not recurrent trap case) under this hypothesis, how we can get rid of it.

PROPOSITION 11 *If* a *isn't a recurrent trap, and if there are no* Φ*-recurrent states, we have for all* j

$$\int_0^\infty \rho_j^a(t)\,dt < \infty \quad . \tag{1}$$

Proof: We shall get (1) by exhausting 5 cases:

1) a isn't recurrent, $j \rightsquigarrow a$;
2) a isn't recurrent, $j \not\rightsquigarrow a$, $j \not\rightsquigarrow B$;
3) a isn't recurrent, $j \not\rightsquigarrow a$, $j \rightsquigarrow B\setminus\{a\}$;
4) a isn't a trap, $j \not\rightsquigarrow B\setminus\{a\}$;
5) a isn't a trap, $j \not\rightsquigarrow B\setminus\{a\}$.

Case 1. j cannot be recurrent for Π (Lemma 2). Then we apply Lemma 3.

Case 2. Since $j \not\rightsquigarrow B$ and j isn't Φ-recurrent, j isn't Π-recurrent (Lemma 1). Then Lemma 3 again.

Case 3. Since $j \rightsquigarrow B\setminus\{a\}$, we can apply Lemma 4.

Case 4. Same as Case 3.

Case 5. We cannot have $j \rightsquigarrow a$, because $j \rightsquigarrow a$, $a \rightsquigarrow B\setminus\{a\}$ (a isn't a trap) would imply $j \rightsquigarrow B\setminus\{a\}$. Therefore $j \not\rightsquigarrow B$, and we conclude as in Case 2.

Let us start proving Proposition 10: denote by e^a the measure

$$e_j^a = \int_0^\infty \rho_j^a(t)\,dt \quad . \tag{2}$$

This is the potential of an entrance law relative to Π^a, and is finite valued: accordingly, it is purely excessive relative to Π^a, and the same is true relative to $\Phi \leqq \Pi^a$. Therefore Proposition 3 of Chapter I shows the existence of an entrance law $(\eta^a_j(t))$ relative to Φ such that

$$e^a_j = \int_0^\infty \eta^a_j(t)\,dt \quad . \tag{3}$$

On the other hand, let us consider the elementary decomposition formula (Proposition 7), applied to Π^a instead of Π'. We write it as

$$\Pi^a(t) = \Phi(t) + \int_0^t \ell^a(u) \otimes \rho^a(t-u)\,du$$

(note that there is just one boundary point for Π^a). We apply this to the row vector $\rho^a(s)$ on the left, getting

$$\rho^a(s+t) = \rho^a(s)\Phi(t) + \int_0^t <\rho^a(s), \ell^a(u)> \rho^a(t-u)\,du \quad .$$

We integrate over s:

$$e^a - \int_0^t \rho^a(s)\,ds = e^a\Phi(t) + \int_0^t <e^a, \ell^a(u)> \rho^a(t-u)\,du \quad . \tag{4}$$

We now define the important function

$$\sigma^{aa}(u) = <e^a, \ell^a(u)> \quad . \tag{5}$$

In the last integral we replace u by $t-s$, and get

$$e^a - e^a\Phi(t) = \int_0^t [1+\sigma^{aa}(t-s)]\rho^a(s)\,ds \quad . \tag{6}$$

We have $\sigma^{aa}(u+v) = <e^a, \ell^a(u+v)> = <e^a, \Phi(v)\ell^a(u)> = <e^a\Phi(v), \ell^a(u)>$; therefore σ^{aa} is decreasing and right continuous by Fatou. Applying (6) to a state j and time t such that $\rho^a_j(t) > 0$, we get that σ^{aa} is integrable in a neighborhood of 0, and therefore finite for $s > 0$. Applying next the fact that $e^a\Phi(v) \to 0$ as $v \to \infty$, we have that $\sigma^{aa}(\infty) = 0$. We can therefore consider the measure $E^a(ds)$ of Lemma 6, such that

$$(7) \qquad \int_0^t [1 + \sigma^{aa}(t-s)]\, E^a(ds) \;=\; 1 \text{ a.e.}$$

The first member of (6) is also equal to $\int_0^t \eta^a(s)\, ds$, from the definition of η^a.

Taking Laplace transforms, for instance, we have at once that

$$(8) \qquad \rho^a(t) \;=\; \int_0^t E^a(ds)\, \eta^a(t-s) \quad \text{for almost every } t.$$

We shall not stop to prove that it holds for all t now, and we turn to a proof of (8) under different conditions: first, when the auxiliary hypothesis is deleted, and then under the assumption that a is a recurrent trap.

To get rid of the auxiliary hypothesis, we shall use the following method. Let $H(\cdot)$ be any SSTM on $\mathbf{I}$, and assume that $\mathbf{I} = A \cup B$, where A and B are disjoint, and B is *absorbing* for H, i.e.,

$$i \in B,\ i \rightsquigarrow j \implies j \in B.$$

(We shall apply our construction to $H = \Phi$ and $H = \Pi^a$, with $A = \mathbf{I}_t$, $B = \mathbf{I}_r$.) Let us denote by Y the process associated with H. We now build a new process Y' and SSTM H' as follows: the new state space $\mathbf{I}'$ is deduced from $\mathbf{I}$ by "building a tower" over B, that is

$$\mathbf{I}' = A \cup (B \times \mathbf{N})\ .^*$$

The process Y' starting at $i' \in \mathbf{I}'$ is described as follows: we take a Poisson process (N_t) of parameter 1, starting at 0, independent of Y. If $i' = i \in A$, then $Y' = Y$ (starting at i) until the time T when Y hits B, and at that time the Poisson clock begins to run, and Y' climbs one step

*$\mathbf{N}$ is the set of positive integers.

in the tower each time N_t climbs one step. More explicitly

$$Y'_t = Y_t \text{ for } t < T, \qquad Y'_t = (Y_t, N_{t-T}) \quad \text{for} \quad t \geqq T .$$

In the same way, if the initial state $i' \epsilon \mathbf{I}'$ belongs to the tower: $i' = (i, n)$, $i \epsilon B$, we set $Y'_t = (Y_t, n+N_t)$ for all t.

We note by H' the transition matrix function of Y'. One can check that to each entrance law (μ_t) one can associate an entrance law (μ'_t) relative to H' (which is the entrance law of the Y' process as described above, corresponding to the Y process with the entrance law (μ_t)).

Now let us take $H = \Pi$, with $A = \mathbf{I}_t$, $B = \mathbf{I}_r$. Then the minimal process associated with Π' is just Φ', and there is no Φ'-recurrent states. According to the preceding discussion, if a isn't a recurrent trap, we can consider a as a boundary point relative to Π', which isn't a recurrent trap, and write

$$\rho'^{a}_{j'}(t) = \int_0^t E^a(ds)\eta'^{a}_{j'}(t-s) \qquad \text{for almost all } t ,$$

with η'^{a} an entrance law for Φ'. To get the desired η^a we have but to set

$$\eta^a_i(t) = \eta'^{a}_i(t) \quad \text{if} \quad i \epsilon \mathbf{I}_t, \qquad \eta^a_i(t) = \sum_n \eta'^{a}_{(i,n)}(t) \quad \text{if} \quad i \epsilon \mathbf{I}_r .$$

We shall see later on page 77 that η^a is a true entrance law, i.e., it is a finite measure.

Proof of Proposition 10: The Recurrent Trap Case

In that case we shall use Lemma 5. Denote by $\mathbf{I}^a$ the set of all $j \epsilon \mathbf{I}$ such that $a \rightsquigarrow j$. Then Π restricted to $\mathbf{I}^a \times \mathbf{I}^a$ is a STM, and all states in $\mathbf{I}^a$ communicate. According to a classical result, due to Derman in the discrete parameter case, there exists a unique measure on $\mathbf{I}^a$ (up to a multiplicative constant) which is positive, nonzero, and invariant by the restriction of Π to $I^a \times I^a$. Let us choose such a measure, extend it by 0 outside of I^a: since $p_{jk}(t) = 0$ for $j \in I^a$, $k \notin I^a$, the measure e^a thus obtained is invariant by Π.

PROPOSITION 12. *The measure* e^a *just defined is purely excessive relative to* Φ, *and therefore admits a representation* $e^a = \int_0^\infty \eta^a(t)\,dt$, *where* η^a *is an entrance law relative to* Φ.

Proof: Since e^a is positive and invariant by Π, it is excessive relative to Φ. Let us set

$$h^a = \lim_{t \to \infty} e^a \Phi(t) \ , \tag{1}$$

then $\langle h^a, \ell(u) \rangle$ is a constant for $u > 0$; and as before (preceding proof, formula (5))

$$\sigma^{aa}(t) = \langle e^a, \ell^a(t) \rangle \ . \tag{2}$$

We have from the elementary decomposition formula (Proposition 7) applied to Π^a, with the remark that $\rho^a = \xi^a$ since a is a trap,

$$\Pi^a(t) = \Phi(t) + \int_0^t \ell^a(u) \otimes \xi^a(t-u)\,du \ .$$

Let us integrate with respect to e^a on the left, and remark that $e^a = e^a \Pi = e^a \Pi^a$. We get

$$\begin{aligned} e^a &= e^a \Phi(t) + \int_0^t \langle e^a, \ell^a(u) \rangle \xi^a(t-u)\,du \\ &= e^a \Phi(t) + \int_0^t \langle e^a, \ell^a(t-s) \rangle \xi^a(s)\,ds \\ &= e^a \Phi(t) + \int_0^t \sigma^{aa}(t-s) \xi^a(s)\,ds \ . \end{aligned} \tag{3}$$

Just as in the preceding proof, we get that σ^{aa} is decreasing and right continuous, integrable in any finite interval. We also have

$$e^a \geqq \sigma^{aa}(\infty) \int_0^t \xi^a(s)\,ds \;;$$

letting $t \to \infty$ we get $\sigma^{aa}(\infty) = 0$ (Lemma 5). Remark now that $\sigma^{aa}(\infty) \geqq$ $< e^a\Phi(v), \ell^a(\infty)> \;\geqq\; < h^a, \ell^a(u)>$ for all u, v. We have therefore $< h^a, \ell^a(u)> \;=\; 0$; on the other hand, for every $i \in I^a$ we have $i \rightsquigarrow a$ (Lemma 5), and therefore $\ell^a(u) > 0$ for some u. Therefore $h^a = 0$, and e^a is purely excessive with respect to Φ. We may thus write (Proposition 3 of Chapter I)

$$(4) \qquad e^a = \int_0^\infty \eta^a(s)\,ds \qquad (\eta^a \text{ is entrance law w.r. to } \Phi).$$

Let us introduce the measure E^a (unbounded) such that

$$(5) \qquad \int_0^t E^a(ds)\sigma^{aa}(t-s) = 1 \qquad \text{for a.e. } t \text{ (Lemma 6)},$$

and invert relation (3), noting that $e^a - e^a\Phi(t) = \int_0^t \eta^a(s)\,ds$ from (3), we get:

$$\xi^a(t) = \int_0^t E^a(ds)\eta^a(t-s) \qquad \text{for a.e. } t.$$

Proposition 10 is thus completely proved. Before going further, we shall make a summary of notations introduced in this proof, and introduce some new notations.

NOTATIONS. $e^a = \int_0^\infty \rho^a(s)\,ds$ except in the recurrent trap case, $e^a =$ the invariant measure on I^a in the recurrent trap case.

In both cases, $e^a = \int_0^\infty \eta^a(s)\,ds$.

In both cases, $\sigma^{aa}(u) = < e^a, \ell^a(u)>$.

E^a is defined by $\int_0^t E^a(ds)[\delta^a + \sigma^{aa}(t-s)] = 1$ for a.e. t, where $\delta^a = 1$ if a isn't a recurrent trap, $\delta^a = 0$ if a is a recurrent trap.

We are going now to end the proof of Theorem 1.

PROPOSITION 13. 1) *The relation (1) of Theorem 1 is true for all* t.
2) *The integral* $<\eta^a(u), \ell^a(v)>$ *depends only on* $u+v$. *We denote it by* $\theta^{aa}(u+v)$.
3) *The measure* $d\sigma^{aa}$ *is absolutely continuous; we have*

$$\sigma^{aa}(t) = \int_t^\infty \theta^{aa}(s)\,ds = <e^a, \ell^a(t)> = <\eta^a(t), L^a(\infty)> . \quad ^*$$

4) E^a *is absolutely continuous except possibly for an atom at* 0.

Proof: We have for a.e. t

$$\rho_j^a(t) = \int_0^t E^a(ds)\eta_j^a(t-s) \quad .$$

The left side is a continuous function of t by Proposition 1 of Chapter I. On the other hand, we know that $\eta_j^a(\cdot)$ is continuous and has a finite limit at 0, and therefore is locally bounded (Chapter I, Proposition 1). The right side therefore is also a continuous function of t from Lemma 6 and Lebesgue's theorem, and thus equality holds for all t.

Next we prove 2). The important point of the proof (to be used later on) is the fact that we have an entrance and an exit law, both relative to the same semi-group. We have

$$\begin{aligned} <\eta^a(u-w), \ell^a(v+w)> &= <\eta^a(u-w), \Phi(w)\ell^a(v)> \\ &= (\eta^a(u-w)\Phi(w), \ell^a(v)> = <\eta^a(u), \ell^a(v)> . \end{aligned}$$

Therefore $<\eta^a(s), \ell^a(t)> = <\eta^a(t), \ell^a(s)> = \theta^{aa}(s+t)$.

Now we compute in two ways:

---

* Recall that $L_i^a(t) = P_i\{\tau^a \leqq t\} = \int_0^t \ell_i^a(s)\,ds$.

$$\int_t^\infty \theta^{aa}(s)\,ds = \int_0^\infty \theta^{aa}(t+u)\,du = \int_0^\infty <\eta^a(u), \ell^a(t)> du$$

$$(1) \qquad = <e^a, \ell^a(t)> = \sigma^{aa}(t)\ ;$$

$$\int_t^\infty \theta^{aa}(s)\,ds = \int_0^\infty <\eta^a(t), \ell^a(u)> du = <\eta^a(t), L^a(\infty)> .$$

The first one of these formulas shows that $d\sigma^{aa}(t) = -\theta^{aa}(t)\,dt$ is absolutely continuous. Therefore E^a is absolutely continuous except possibly for an atom at 0 if $\sigma^{aa}(0+) < \infty$ (Lemma 6). There is another, more elementary, proof of the continuity of E^a in the nonrecurrent trap case. Let us compute

$$(2) \qquad <\rho^a(t), L^a(\infty)> = <\int_0^t \eta^a(t-s)\,E^a(ds), L^a(\infty)>$$

$$= \int_0^t E^a(ds) <\eta^a(t-s), \int_0^\infty \ell^a(u)\,du>$$

$$= \int_0^t E^a(ds) \int_{t-s}^\infty \theta^{aa}(v)\,dv$$

$$= \int_0^t E^a(ds)\,\sigma^{aa}(t-s) .$$

Now $1 = \int_0^t E^a(ds)[1 + \sigma^{aa}(t-s)]$, and therefore we get at once

$$(3) \qquad <\rho^a(t), L^a(\infty)> = 1 - E^a(t) .$$

Now this is just $E_a[L^a_{X(t)}(\infty)]$, computed on the Π^a-process: it is therefore continuous for $t > 0$ since $X(\cdot)$ is stochastically continuous and $L^a_\cdot(\infty) \leqq 1$. In the recurrent trap case, the first member of (2) is equal to 1, and we get no information about E^a.

The Complete Decomposition Formula

Recall our main purpose (ends of Sections 1 and 2): we want to express the semi-group Π using only quantities which are related to the minimal semi-group Φ. Proposition 10 will allow us to fulfill this wish, and without need of the more precise parts of the statement of Theorem 1. To give a complete statement, we write again some equations previously written.

THEOREM 2. 1) *We have* ($\hat{}$ *denoting Laplace transforms)*

$$\hat{\Pi}(\lambda) = \hat{\Phi}(\lambda) + \sum_{a \in E} \hat{\ell}^{a}(\lambda) \oplus \hat{\xi}^{a}(\lambda) \quad .$$

The sum being taken over the set of all passable exits.
2) *Let* B *be the set of all passable boundary points. Let* $\hat{F}$ *be the matrix on* B *with the coefficients*

$$\hat{F}_{ab}(\lambda) = \int_0^\infty F^{ab}(ds)e^{-\lambda s} \quad ;$$

and let $\hat{E}$ *be the diagonal matrix on* B *with diagonal coefficients:*

$$\hat{E}_{aa}(\lambda) = \int_0^\infty E^{a}(ds)e^{-\lambda s} \quad .$$

Denote by $\hat{\eta}_j(\lambda), \hat{\xi}_j(\lambda)$, *for each* $j \in I$, *the column vector on* B *with coefficients* $\hat{\eta}_j^a(\lambda), \hat{\xi}_j^a(\lambda)$. *Then* $I - \hat{F}(\lambda)$ *is invertible and we have*

$$(2) \qquad \hat{\xi}_j(\lambda) = [I - \hat{F}(\lambda)]^{-1} \hat{E}(\lambda) \hat{\eta}_j(\lambda) \quad .$$

Theorem 2 appeared in [3] in 1966. A decomposition of this form was first given by Feller (1957) under the hypothesis that both the exit and entrance boundaries are finite. David Williams (1964) extended the result to the present case without mention of entrance boundary. Dynkin (1967) gave another derivation with an entrance boundary. These authors were concerned with the construction of semi-groups without studying the underlying stochastic processes.

CHAPTER IV

PROBABILISTIC INTERPRETATIONS AND ADDITIONAL RESULTS

Section 1 gives intrinsic meanings to the basic quantities introduced in Chapter III. Section 2 contains a useful complement to Theorem 1, the analytic criterion for stickiness, and a related analytic result. Section 3 deals with the "identification problem": how can we compute analytically basic quantities whose definitions are probabilistic?

Section 1. Evaluation of Various Probabilities

We are going to show now that all the quantities we have introduced have simple probabilistic meanings. This will imply that the decompositions we have given are "natural". However, there will be a sharp difference between the not recurrent trap and the recurrent trap case: in the first case, e^a was defined in an intrinsic way, and E^a was a probability measure. On the other hand, in the second case, e^a was defined up to a multiplicative constant, the same being therefore true for $\eta^a(\cdot)$, and therefore for E^a (which is unbounded): if η^a is replaced by $c\eta^a$, E^a is replaced by E^a/c. This explains why our intrinsic interpretations will be valid only in the not recurrent trap case.

First, we shall generalize Proposition 13 of the last chapter.

DEFINITION. *For any two boundary points* a, b *we set*

(1) $$\sigma^{ab}(t) = \langle e^a, \ell^b(t)\rangle , \qquad t > 0 .$$

This is a right continuous and decreasing function.

PROPOSITION 1. 1) $<\eta^a(t), \ell^b(s)>$ *depends only on* $t+s$. *We shall denote it by* $\theta^{ab}(t+s)$. *We have*

$$(2) \qquad \sigma^{ab}(t) = \int_t^\infty \theta^{ab}(s)ds = <e^a, \ell^b(t)> = <\eta^a(t), L^b(\infty)> .$$

Proof: See Proposition 13 of Chapter III.

We start with the interpretation of E^a in the not recurrent trap case.

DEFINITION. *The random variable* γ^a *is defined by*

$$\gamma^a(\omega) = \sup\{\tilde{S}_a(\omega) \cap]0, \beta^a(\omega)]\} \; ; \; ^*$$

γ^a is the "last" time the value a is taken before switching. It is a r.v., but not an optional one. Of course, if a is a recurrent trap, γ^a is a.s. equal to ∞.

Proof: $P_a\{\gamma^a > t\} = P_a\{\beta^a > t, X(\tau_{(t)}-) = a\}$ (we express that: switching has not yet taken place at t, and that the first boundary value after t is still a). Applying the Markov property we get

$$P_a\{\gamma^a > t\} = <\rho^a(t), L^a(\infty)> .$$

This is equal to $1 - E^a(t)$, from the computation at the end of the proof of Proposition 13, Chapter III.

We are going to give a more precise result, involving the joint distribution of γ^a, β^a, and the switching value from a. It will be useful to introduce a notation, Z^a, for this switching value, and to recall the precise definition we gave for it in Chapter III, Section 2 (page 53). $Z^a = b$ means that $\beta^a < \infty$ and that

either b is nonsticky and $\beta^a \in \tilde{S}_b$; or

b is sticky, $\beta^a \notin \tilde{S}_c$ for all nonsticky $c \neq a$, and $\beta^a \in S_b^+$.

---

* As usual, the sup of the empty set in T is 0.

PROPOSITION 3. *For* $a \neq b$, $s < t$ (a *not a recurrent trap*),

$$(1) \qquad P_a\{\gamma^a < \beta^a,\ \gamma^a \in ds,\ \beta^a \in dt,\ Z^a = b\} = E^a(ds)\theta^{ab}(t-s)dt \ ;$$

or, in integrated form

$$(2) \qquad P_a\{\gamma^a < s < t < \beta^a < t',\ Z^a = b\} = \int_0^s E^a(du)\int_t^{t'} \theta^{ab}(v-u)dv \ .$$

Proof: Let us compute the left side of (2). We must express: that we have not yet switched from a at time s, that between s and t' there is a first hit of the boundary taking place at b, but that this hit has not yet occurred at time t. Therefore we get the value

$$\begin{aligned}
\langle\rho^a(s), L^b(t'-s) - L^b(t-s)\rangle &= \int_0^s E^a(du)\langle\eta^a(s-u), L^b(t'-s) - L^b(t-s)\rangle \\
&= \int_0^s E^a(du)\langle\eta^a(s-u), \int_{t-s}^{t'-s} \ell^b(w)\,dw\rangle \\
&= \int_0^s E^a(du)\int_{t-s}^{t'-s} \theta^{ab}(s+w-u)dw \ ,
\end{aligned}$$

which is the right side of (2). Integrating over t, and noting that $\int_0^\infty \theta^{ab}(u)\,du = \sigma^{ab}(0)$, we get

COROLLARY. $P_a\{\gamma^a < \beta^a,\ Z^a = b \mid \gamma^a\} = P_a\{\gamma^a < \beta^a,\ Z^a = b\} = \sigma^{ab}(0)$. (To get the integrated form, remark that γ^a is a.s. finite if a isn't a recurrent trap: Proposition 2, for instance.)

PROPOSITION 4. $P_a\{Z^a = b \mid \gamma^a\} = P_a\{Z^a = b\} = F^{ab}(\infty)$.

Proof: Let us compute $P_a\{t < \gamma^a,\ Z^a = b\}$. We must express that at t we haven't yet switched from a, that after t we have still a first boundary hit taking place at a, and that after this first hit $\tau^a_{(t)}$ the first switching value is b. Applying the strong Markov property at $\tau^a_{(t)}$ we get

$$P_a\{t < \gamma^a, Z^a = b\} = <\rho^a(t), L^a(\infty)> F^{ab}(\infty)$$

$$= P_a\{t < \gamma^a\} F^{ab}(\infty) \qquad \text{(Proposition 2)}$$

$$(= P_a\{t < \gamma^a\} P_a\{Z^a = b\}) \; .$$

COROLLARY. $P_a\{\gamma^a = \beta^a, Z^a = b \mid \gamma^a\} = P_a\{\gamma^a = \beta^a; Z^a = b\} = F^{ab}(\infty) - \sigma^{ab}(0)$. (Note that it is far from obvious that the right member is $\geqq 0$!) This event can happen only for a *sticky* b, because when b is nonsticky $Z^a = b$ implies that β^a is the right endpoint of an interval in which $X(\cdot) \in I$, and therefore we can't have $\beta^a = \gamma^a$. On the other hand, it can be shown that $\gamma^a = \beta^a$, $Z^a = b$ can happen when a and b both are sticky.

PROPOSITION 5. $P_a\{\beta^a = \infty \mid \gamma^a\} = P_a\{\beta^a = \infty\} = \lim_{t \to \infty} <\rho^a(t), 1>$ (not recurrent trap case).

Proof: We have $<\rho^a(t), 1> = P_a\{\beta^a > t\}$: that gives the second equality. On the other hand

$$1 = P_a\{\beta^a = \infty \mid \gamma^a\} + \sum_{b \neq a} P_a\{\gamma^a < \beta^a, Z^a = b \mid \gamma^a\} + \sum_{b \neq a} P_a\{\gamma^a = \beta^a, Z^a = b \mid \gamma$$

The left side, and all terms in the sums on the right side, are a.s. constants. The same is therefore true for $P_a\{\beta^a = \infty \mid \gamma^a\}$.

Note that in the recurrent trap case the integrated from remains valid (and trivial), both sides being equal to 1.

NOTATION. We set $\lim_{t \to \infty} <\rho^a(t), 1> = c^a$.

PROPOSITION 6. *Set* $\tilde{c}^a = \lim_{t \to \infty} <\eta^a(t), 1>$. *Then*

1) $<\eta^a(t), 1 - L(\infty)> = \tilde{c}^a$ *for all* t;*

---

* We set $L_i(t) = P_i\{\tau \leq t\}$.

2) *in the not recurrent trap case,* $\tilde{c}^a = c^a$. *In the recurrent trap case* $\tilde{c}^a = 0$ *(while* $c^a = 1$*):* $\tilde{c}^a = \delta^a c^a$!

3) $<\eta^a(t), 1>$ *is finite for all* $t > 0$, *and*

$$\lim_{t \to 0} <\eta^a(t), 1> \; = \; \tilde{c}^a + \Sigma_b \; \sigma^{ab}(0+) \quad .$$

Proof: The function $<\eta^a(.), 1>$ is decreasing, since $\eta^a(.)$ is a Φ-entrance law, and 1 is Φ-excessive. This implies at once the existence of $\tilde{c}^a$ (but not yet its finiteness). If we had $<\eta^a(t), 1> \; = \; \infty$ for some $t > 0$, we would have $<\eta^a(t-s), 1> \; = \; \infty$ for $s < t$. On the other hand

$$1 \geqq \; <\rho^a(t), 1> \; = \int_0^t E^a(ds)<\eta^a(t-s), \; 1> \quad .$$

Our property would therefore imply $E^a(t) = 0$, but this would contradict $1 = \int_0^t E^a(ds)[\sigma^{aa}(t-s) + \delta^a]$. The first part of 3) is proved.

Next, we write that $\tilde{c}^a = \lim_{s \to \infty} <\eta^a(t+s), 1> \; = \lim_{s \to \infty} <\eta^a(t), \Phi(s), 1>$ $= \; <\eta^a(t), 1 - L(\infty)>$. (We have used the fact that η^a is an entrance law, the fact that $\Phi(s)1$ decreases to $1 - L(\infty)$ as $s \to \infty$, and Lebesgue's theorem justified by the finiteness argument above.)

We now compute $\tilde{c}^a$. If $\delta^a = 0$ (recurrent trap case), we write it as $\Sigma_j \; \eta_j^a(t)[1 - L_j(\infty)]$. This sum over $\mathbf{I}$ can be restricted to $\mathbf{I}^a$, since $\eta_j^a(\cdot) = 0$ for $j \notin \mathbf{I}^a$. On the other hand if $j \in \mathbf{I}^a$ we have $L_j^a(\infty) = 1$, and we get that $\tilde{c}^a = 0$. If $\delta^a = 1$ (not recurrent trap case), let us write $<\rho^a(t), 1> \; = \int_0^t E^a(ds)<\eta^a(t-s), 1>$ and let $t \to \infty$. The left side tends to c^a; on the right side, the integrand converges pointwise to $\tilde{c}^a$. Since E^a has a total mass equal to 1, we get from Fatou's lemma that $\tilde{c}^a \leqq c^a$. If $<\eta^a(t), 1>$ remains bounded as $t \to 0$, we deduce from Lebesgue's theorem that $\tilde{c}^a = c^a$. The general case is postponed until after the corollary.

Finally, let us prove the last part of 3): we have

$$\begin{aligned} <\eta^a(t), 1> \; &= \; <\eta^a(t), 1 - L(\infty)> + <\eta^a(t), L(\infty)> \\ &= \; <\eta^a(t), 1 - L(\infty)> + \sum_b <\eta^a(t), L^b(\infty)> \; = \end{aligned}$$

$$= \langle \eta^a(t), 1 - L(\infty) \rangle + \sum_b \sigma^{ab}(t)$$

(see Proposition 1). Now we just let $t \to 0$.

COROLLARY. $\langle \eta^a(t), 1 \rangle - \sigma^{aa}(t) \leqq 1$ *for all* t.

Proof: This follows at once from the preceding computation, since the difference is $\tilde{c}^a + \Sigma_{b \neq a} \sigma^{ab}(t)$, and we know that $\tilde{c}^a \leq c^a$, $\sigma^{ab}(t) \leqq \sigma^{ab}(0) \leqq F^{ab}(\infty)$, and $c^a + \Sigma_{b \neq a} F^{ab}(\infty) = 1$ * from the probabilistic interpretation of c^a as $P_a\{\beta^a = \infty\}$.

End of the Proof that $\tilde{c}^a = c^a$ *in the Not Recurrent Trap Case:* Let us set

$$f_n(s) = \begin{cases} \langle \eta^a(n-s), 1 \rangle & \text{if } s \leqq n; \\ 0 & \text{if } s > n. \end{cases} \qquad g_n(s) = \begin{cases} \sigma^{aa}(n-s) & \text{if } s \leqq n \\ \sigma & \text{if } s > n \end{cases}$$

As $n \to \infty$, $f_n(s) \to \tilde{c}^a$, $g_n(s) \to 0$ (pointwise), and the corollary implies that $|f_n - g_n| \leqq 1$. Our problem consists in showing that $\int_0^\infty E^a(ds) f_n(s)$ tends to $\tilde{c}^a$: otherwise stated, that the f_n are uniformly integrable with respect to E^a. Equivalently, we must show that the g_n are uniformly integrable, and this in turn is equivalent to the fact that $\int_0^\infty E^a(ds) g_n(s)$ tend to 0 as $n \to \infty$. This is obvious, since $\int_0^n E^a(ds)[1 + \sigma^{aa}(n-s)] = 1$;

$$\int_0^\infty E^a(ds) g_n(s) = 1 - E^a(n) \xrightarrow[n \to \infty]{} 0 .$$

We end this section with a computation of the quantity $F^{ab}(\infty) - \sigma^{ab}(0)$.

PROPOSITION 7.

$$F^{ab}(\infty) - \sigma^{ab}(0) = \lim_{t \to 0} \frac{F^{ab}(t)}{E^a(t)} \qquad (a \neq b) .$$

---

* $\leqq 1$ in the substochastic case.

Proof: Of course everything is true and trivial in the recurrent trap case! We have in the not recurrent trap case

$$(1) \qquad F^{ab}(\infty) - F^{ab}(t) = \langle \rho^a(t), L^b(\infty) \rangle + \langle \rho^a(t), L^a(\infty) \rangle F^{ab}(\infty) \ .$$

(The left side is the probability, starting at a, that no switching has occurred at time t, and that switching to b occurs after t. If we split this event in two, according to whether b is the first boundary hit after t, or whether this first boundary is a and the switching occurs after $\tau^a_{(t)}$, and apply the strong Markov property to the second term, then we get the right side of (1).)

Now we replace $\langle \rho^a(t), L^a(\infty) \rangle$ by $1 - E^a(t)$ (Proposition 2) and $\langle \rho^a(t), L^b(\infty) \rangle$ by $\int_0^t E^a(ds)\sigma^{ab}(t-s)$ (same computation as in the proof of Proposition 13, p. 71). Therefore

$$(2) \qquad E^a(t)F^{ab}(\infty) = F^{ab}(t) + \int_0^t E^a(ds)\sigma^{ab}(t-s) \ .$$

Remark that $E^a(t) > 0$ for all t, since $\int_0^t [1 + \sigma^{aa}(t-s)]\, E^a(ds) = 1$. Dividing by $E^a(t)$ and letting $t \to 0$ we get the result at once.

Section 2. A Criterion for Stickiness

THEOREM 1. *The following properties are equivalent:*

1) a *is sticky;*
2) $\sigma^{aa}(0+) = +\infty$ *;*
3) $E^a(\{0\}) = 0$ *;*
r) $\lim_{t \to 0} \langle \eta^a(t), 1 \rangle = +\infty$.

The fact that 2) and 3) are equivalent has nothing to do with boundaries: it was stated in Lemma 6 of the last chapter. The fact that 2) and 4) are equivalent is an obvious consequence of the corollary to Proposition 6. Before we give the proof of the interesting part (equivalence to 1)), we give one notation, and a little computation:

NOTATION. We set for $i \in I$

$$K_i^a(t) = P_i\{\tilde{S}_a \cap]0,t[\neq \emptyset\} .$$

We obviously have $L_i^a(t) \leqq K_i^a(t) \quad (\leqq L_i(t))$.

We shall need the following computation.

LEMMA. $<\rho^a(u), L^b(v)> = \int_0^u E^a(ds)[\sigma^{ab}(u-s) - \sigma^{ab}(v+u-s)]$.

Indeed, the first member is equal to

$$< \int_0^u E^a(ds)\eta^a(u-s), \int_0^v \ell^b(t)dt >$$

$$= \int_0^u E^a(ds) <\eta^a(u-s), \int_0^v \ell^b(t)dt>$$

$$= \int_0^u E^a(ds) \int_0^v \theta^{ab}(u-s+t)dt$$

$$= \int_0^u E^a(ds) \left[\int_{u-s}^\infty \theta^{ab}(w)\,dw - \int_{v+u-s}^\infty \theta^{ab}(w)\,dw\right] .$$

We conclude with Proposition 1.

Proof: $E^a(\{0\}) = 0$ IMPLIES THAT a IS STICKY.

We want to prove that for every $t > 0$, $P_a\{\tilde{S}_a \cap]\varepsilon,t[\neq \emptyset\} \to 1$ as $\varepsilon \to 0$. Now this is $<\xi^a(\varepsilon), K^a(t-\varepsilon)>$, greater than $<\rho^a(\varepsilon), L^a(t-\varepsilon)>$ (since $\rho^a \leq \xi^a$ and $L^a \leq K^a$). Let us show that this tends to 1. We compute it, using the lemma, as

$$\int_0^\varepsilon E^a(ds)[\sigma^{aa}(\varepsilon-s) - \sigma^{aa}(t-s)]$$

$$= 1-\delta^a E^a(\varepsilon) - \int_0^\varepsilon E^a(ds)\sigma^{aa}(t-s) \geqq 1-\delta^a E^a(\varepsilon) - E^a(\varepsilon)\sigma^{aa}(t-\varepsilon) .$$

(We have used here the relation $1 = \int_0^\varepsilon E^a(ds)[\delta^a + \sigma^{aa}(\varepsilon-s)]$ and the fact that σ^{aa} ia decreasing.) Now we let $\varepsilon \to 0$, and just remark that $E^a(\varepsilon) \to 0$ if E^a has no atom at 0.

Proof: $\sigma^{aa}(0) < \infty$ IMPLIES THAT a IS NOT STICKY.

The simplest proof seems to be the following: first we begin with the case of a (substochastic) Π with just one boundary point a —this is the first time we shall really need a *substochastic* Π! Then we want to show that

$$\lim_{t\to 0}\lim_{\varepsilon\to 0} P_a\{\tilde{S}_a \cap]\varepsilon, t[\neq \emptyset\} = \lim_{t\to 0}\lim_{\varepsilon\to 0} <\xi^a(\varepsilon), K^a(t-\varepsilon)> = 0 \ .$$

But in this case $\xi^a = \rho^a$ and $K^a = L^a$, therefore, using the lemma above we are reduced to showing that

$$\lim_{t\to 0}\lim_{\varepsilon\to 0}\int_0^\varepsilon E^a(ds)[\sigma^{aa}(\varepsilon-s)-\sigma^{aa}(t-s)] = 0 \ .$$

Now since σ^{aa} is bounded the limit as $\varepsilon\to 0$ is $E^a(\{0\})[\sigma^{aa}(0)-\sigma^{aa}(t)]$, which tends to 0 as $t\to 0$.

To deal with the general case, we replace Π by Π^a: this is a substochastic transition matrix with just a as boundary point, and obviously the minimal semi-group, ρ^a, e^a, and therefore E^a, σ^{aa}, are the same for both semi-groups. Therefore from the above a is nonsticky for Π^a, and the probabilistic interpretation implies the same result for Π.

PROPOSITION 8. *The following limits exist and are finite, for any two boundary points* a, b *(distinct or not) and any* $t > 0$ *(t may be* $+\infty$*).*

(1) $$\check{L}^{ab}(t) = \lim_{s\to 0} <\rho^a(s), L^b(t-s)> \ ;$$

(2) $$L^{ab}(t) = \lim_{s\to 0} <\xi^a(s), L^b(t-s)> \ .$$

We have $L^{ab}(t) = \check{L}^{ab}(t) + F^{ab}(\{0\})$ (hence they are equal if a is sticky); *and*

$$L^{ab}(t) = \check{L}^{ab}(t) = \delta^{ab} \text{ (Kronecker symbol) } \textit{if a is sticky,}$$

$$\check{L}^{ab}(t) = E^a(\{0\})[\sigma^{ab}(0)-\sigma^{ab}(t)] \textit{ if a is nonsticky.}$$

(In (1) and (2), $L^b(t-s)$ can be replaced by $L^b(t)$.)

Proof: We have, according to the lemma in the proof of Theorem 1,

$$<\rho^a(s), L^b(t-s)> = \int_0^s E^a(du)[\sigma^{ab}(s-u) - \sigma^{ab}(t-u)] \ .$$

This shows the existence of $\tilde{L}^{ab}(t)$: 1) if $a \neq b$ (because then $\sigma^{ab}(0) \leqq F^{ab}(\infty)$; 2) if $a = b$ and a is nonsticky (then $\sigma^{aa}(0) < \infty$). We have then

$$\tilde{L}^{ab}(t) = E^a(\{+\})[\sigma^{ab}(0) - \sigma^{ab}(t)] \ .$$

If a is sticky, $E^a(\{0\}) = 0$, and $\tilde{L}^{ab}(t) = 0$ if $a \neq b$.

We now deal with the case where a is sticky, $b = a$. Then, using a computation already given in the proof of Theorem 1,

$$<\rho^a(s), L^a(t-s)> = 1 - \delta^a E^a(s) - \int_0^s E^a(du)\sigma^{aa}(t-u) \ .$$

Since a is sticky, $E^a(s) \to 0$ as $s \to 0$, and the above tends to 1 (remember that $t > 0$!). All problems concerning $\tilde{L}^{ab}$ are thus settled.

To deal with L^{ab}, we use Proposition 8 of Chapter III:

$$<\xi^a(s), L^b(t-s)> = <\rho^a(s), L^b(t-s)> + \sum_{c \neq a} \int_0^s F^{ac}(du)<\xi^c(s-u), L^b(t-s)>$$

Since the boundary is finite, and $<\xi^c(s-u), L^b(t-s)> \leqq 1$, the total contribution of $]0, s[$ in the integrals tend to 0 with s, whether or not a is sticky. If now we assume that a is sticky, we have $F^{ac}(0) = 0$ for every $c \neq a$, and therefore the contribution of $\{0\}$ is also 0. Hence if a is sticky, $L^{ab}(t) = \tilde{L}^{ab}(t)$ ($= \delta^{ab}$). Assume now that a is nonsticky. The only c's which contribute to the sum are those such that $F^{ac}(0) > 0$, but they must be sticky, and therefore we know from the discussion we just have made that $<\xi^c(s), L^b(t-s)>$ tends to δ^{cb}! Therefore $L^{ab}(t)$ exists and is equal to

$$\tilde{L}^{ab}(t) + \sum_{\substack{c \text{ sticky} \\ c \neq a}} F^{ac}(0)\delta^{cb} .$$

The proposition follows at once.

The following proposition (which will be useful in the next section) sharpens Proposition 8.

PROPOSITION 9. *For every* $a \in \mathbf{B}$, *every* $t > 0$ (t *may be* $+\infty$), *we have as* $s \to 0$

$$(1) \qquad 1 - \langle \xi^a(s), L^a(t) \rangle \sim E^a(s)[\delta^a + \sigma^{aa}(t)] \; ;^*$$

and, if $b \neq a$ *and* $\delta^a = 1$ *(not recurrent trap case),*

$$(2) \qquad \langle \xi^a(s), L^b(t) \rangle \sim E^a(s)[F^{ab}(\infty) - \sigma^{ab}(t)] \; .$$

Proof: We have

$$\langle \xi^a(s), L^a(t) \rangle = 1 - \langle \rho^a(s), L^a(t) \rangle - \sum_{b \neq a} \int_0^s F^{ab}(du) \langle \xi^b(s-u), L^a(t) \rangle \; .$$

We shall first show that every term in the sum is $o(E^a(s))$. First of all, this is obvious if $F^{ab}(s)/E^a(s) \to 0$, since the bracket $\langle \; \rangle$ is at most 1. Now the limit of this ratio is $F^{ab}(\infty) - \sigma^{ab}(0)$ (Proposition 7), which is 0 if b is nonsticky (see corollary to Proposition 4). Therefore nonsticky b's are ruled out. If b is sticky, on the other hand, we know that

$$\langle \xi^b(s-u), L^a(t) \rangle \leqq \langle \xi^b(s-u), L^a(2t-s+u) \rangle$$

for s small enough, is thus majorized, as $s \to 0$, by a quantity which tends to $L^{ba}(2t) = 0$ (Proposition 8), Therefore the integral is $o(F^{ab}(s))$, and $o(E^a(s))$ since $F^{ab}(s) = 0(E^a(s))$.

To prove (1), we must therefore show that

$$1 - \langle \rho^a(s), L^a(t) \rangle \sim E^a(s)[\delta^a + \sigma^{aa}(t)] \; .$$

---

* This is slightly incorrect if $t = \infty$, $\delta^a = 0$, and should be written then $1 - \langle \xi^a(s), L^a(\infty) \rangle = o(E^a(s))$.

But the first term is equal to

$$1-\int_0^s E^a(du)[\sigma^{aa}(s-u)-\sigma^{aa}(s+t-u)] = \delta^a E^a(s)-\int_0^s E^a(du)\sigma^{aa}(t+s-u$$

and (1) follows at once.

We now come to (2). We have as before

$$<\xi^a(s), L^b(t)> = <\rho^a(s), L^b(t)> + \sum_{c \neq a} \int_0^s F^{ac}(du)<\xi^c(s-u), L^b(t)>$$

We can see as above that all terms in the sum are $o(E^a(s))$ for nonsticky c's, or for sticky c's $\neq$ b. But if b is sticky we cannot neglect

$$\int_0^s F^{ab}(u)<\xi^b(s-u), L^b(t)> \quad .$$

This is equivalent to $F^{ab}(s)$ (Proposition 8), and therefore also to $E^a(s)[F^{ab}(\infty) - \sigma^{ab}(0)]$ (Proposition 7). On the other hand,

$$<\rho^a(s), L^b(t)> = \int_0^s E^a(du)[\sigma^{ab}(s-u) - \sigma^{ab}(s+t-u)]$$

is equivalent to $E^a(s)[\sigma^{ab}(0) - \sigma^{ab}(t)]$, and we get the result by additior

Section 3. The Identification Problem

Most of the basic quantities we have introduced are defined from prob bilistic considerations, and it is natural to ask whether we may define the analytically, without reference to the processes. This question is answer in this section. Of course, these analytical definitions may involve multi passages to the limit, and be quite intractable from the computational poi of view.

We start with the STM Π—that is, with the functions $p_{ij}(\cdot)$. Then we may compute successively

- the derivatives q_i, q_{ij} (and therefore the transition matrix of the ju chain);

– the functions $f_{ij}(\cdot)$, as minimal solutions of the Kolmogorov system. An explicit calculation is given, for instance in [1 ; § II.18].

Now an awkward step: we must find the number of exits, and associate to each exit a the corresponding functions $L^a_\cdot(\infty)$, $L^a_\cdot(t)$, $\ell^a_\cdot(t)$. In fact, all can be deduced from the first one, since

$$L^a(t) = L^a(\infty) - \Phi(t)\, L^a(\infty) ; \qquad \ell^a(t) = -\frac{d}{dt} L^a(t) .$$

The simplest analytic (but not computational!) definition consists in saying that the functions $L^a_\cdot(\infty)$ are the nonzero extremal points of the set of all positive, harmonic and purely excessive functions relative to the jump matrix (see Proposition 9 of Chapter II). The standard way of obtaining them "explicitly" is the Martin procedure applied to the jump chain.

Next, an even more awkward step: we must define without probability the entrance laws $\xi^a(\cdot)$ (this will imply the knowledge of the boundary, since a boundary point is a class of indistinguishable exits). The only analytic formula to achieve this is the following, involving Laplace transforms (Chapter III, Proposition 7):

$$\sum_{a \in E} \hat{\ell}_i^a(\lambda)\, \hat{\xi}_j^a(\lambda) = \hat{p}_{ij}(\lambda) - \hat{f}_{ij}(\lambda) .$$

For each j, this is an overdetermined system of (an infinity of) linear equations for the quantities $\hat{\xi}_j^a(\lambda)$.

Now begins the more interesting part of the discussion. We choose a boundary *point* a, and we want to construct ρ^a, F^{ab}, σ^{ab}, E^a, η^a. The first step obviously is:

To recognize whether a *is a recurrent trap or not.* The criterion is: a is a recurrent trap if and only if $<\xi^a(s), L^a(\infty)> = 1$, $\forall s > 0$ (Proposition 9). Then we split the discussion in two. We begin with the simpler case.

If a *is a recurrent trap.* Since we know ξ^a, we know I^a, and we can compute e^a: for instance, we fix some state k in I^a, and denote by ${}_kp_{ij}(t)$ transition probabilities with the "taboo" k. Then a choice of e^a is

$$e_j^a = \int_0^\infty {}_k p_{kj}(s)\,ds \qquad \text{(see [1; p.222]).}$$

Next,

$$\eta^a(t) = \frac{d}{dt} e^a[I - \Phi(t)] \ .$$

Then

$$\sigma^{aa}(t) = \langle \eta^a(t), L^a(\infty) \rangle \qquad \text{(Proposition 1)}$$

Then $E^a(\cdot)$ is the unique solution of $1 = \int_0^t E^a(ds)\sigma^{aa}(t-s)$.

And this is all, since $\rho^a = \xi^a$, and F^{ab}, σ^{ab} are 0 for all $b \neq a$.

If a *is not a recurrent trap*. We first get σ^{aa} from the formula (Proposition 9)

$$1 + \sigma^{aa}(t) = \lim_{s \to 0} \frac{1 - \langle \xi^a(s), L^a(t) \rangle}{1 - \langle \xi^a(s), L^a(\infty) \rangle} \ ;$$

then E^a as the unique solution of the integral equation

$$1 = \int_0^t E^a(ds)[1 + \sigma^{aa}(t-s)] \ ;$$

then $F^{ab}(\infty)$ by the formula (2) of Proposition 9, with $t = \infty$,

$$F^{ab}(\infty) = \lim_{s \to 0} \frac{\langle \xi^a(s), L^b(\infty) \rangle}{1 - \langle \xi^a(s), L^a(\infty) \rangle} \ ;$$

and $\sigma^{ab}(t)$ from the same formula, with t arbitrary,

$$\sigma^{ab}(t) = F^{ab}(\infty) - \lim_{s \to 0} \frac{\langle \xi^a(s), L^b(t) \rangle}{1 - \langle \xi^a(s), L^a(\infty) \rangle} \ .$$

We get $F^{ab}(t)$ from formula (2) of Section 1 (Proposition 7),

$$F^{ab}(t) = \int_0^t E^a(ds)[F^{ab}(\infty) - \sigma^{ab}(t-s)] \ .$$

Next, ρ^a is computable:

$$\rho^a(t) = \xi^a(t) - \sum_{b \neq a} \int_0^t F^{ab}(ds)\xi^b(t-s) ,$$

and finally we get η^a either by the formulas

$$e^a = \int_0^\infty \rho^a(t)\,dt, \qquad \eta^a = \frac{d}{dt} e^a[I - \Phi(t)] ,$$

if there are no Φ-recurrent states, or in the general case by the Laplace transform formula

$$\hat{\rho}^a(\lambda) = \hat{E}^a(\lambda)\hat{\eta}^a(\lambda) .$$

An alternative and suggestive formula for η^a in case $\delta^a = 1$ is as follows:

$$\eta^a(t) = \lim_{s \downarrow 0} \frac{\xi^a(s)\Phi(t-s)}{1 - \langle \xi^a(s), L^a(\infty)\rangle} .$$

We leave this as an exercise.

Finally, a description of the evolution in time of the sample functions of the process, in relation to the boundary, can be found on pp. 161-2 of [3].

APPENDIX

PROOF OF LEMMA 6, CHAPTER III, SECTION 3

According to a theorem due to Lévy, given c and σ as stated, there exists a stochastic process $(Y_s)_{s \geqq 0}$ with stationary independent increments (defined on some probability space Ω) with values in the interval $[0, \infty[$, such that

1) $E[e^{-\lambda Y_s}] = \exp(-cs + s \int_0^\infty (1 - e^{-\lambda t})\, d\sigma(t)) \qquad (\lambda \geqq 0,\ s \geqq 0)$;

2) for every $\omega \in \Omega$, the sample function $Y(\omega)$ is right continuous and increasing.

Condition 1) implies that $Y_0 = 0$. For our purpose, condition 2) might be replaced by stochastic continuity. Denote by η_s the distribution of Y_s, and set

$$E(\cdot) = \int_0^\infty \eta_s(\cdot)\, ds.$$

The existence of the integral is a consequence of condition 2); E is a completely additive set function, but we don't know yet whether it has any finiteness properties. An easy interchange of integration yields

$$\int_0^\infty e^{-\lambda s}\, dE(s) = \int_0^\infty ds \int_0^\infty e^{-\lambda t}\, d\eta_s(t)$$

$$= \frac{1}{c - \int_0^\infty (1 - e^{-\lambda t})\, d\sigma(t)} = \frac{1}{c + \lambda \hat{\sigma}(\lambda)} .$$

Since the last member is finite for $\lambda > 0$, E is a measure. We deduce that

E is a solution of the equation

$$\hat{E}(\lambda)[c + \lambda\hat{\sigma}(\lambda)] = 1 \quad . \tag{3}$$

Inverting the Laplace transform, we get that for *almost every* t

$$\int_0^t [c + \sigma(t-s)]\,E(ds) = 1 \quad . \tag{4}$$

If $\sigma(0+) < \infty$, the left side in (4) is right continuous in t, hence (4) holds for all $t > 0$. The only remaining nontrivial assertion in Lemma 6 is part 3)

Suppose then $\sigma(t) = \int_t^\infty \theta(s)\,ds$ for some positive Borel function θ. Decompose θ into $\theta_1 = \theta I_{[0,\varepsilon[}$, $\theta_2 = \theta I_{[\varepsilon,\infty]}$, and associate to these two functions the decreasing functions σ_1, σ_2, the processes Y_s^1, Y_s^2, and their distributions η_s^1, η_s^2 as above, but construct the processes in such a way that they are *independent*. It is very easy to check now that $Y_s = Y_s^1 + Y_s^2$ is a process with independent increments satisfying 1) and 2): therefore we have, with the above notations

$$\eta_s = \eta_s^1 * \eta_s^2 \quad .$$

On the other hand, put $C = \int_\varepsilon^\infty \theta(t)\,dt$ and $h(t) = \theta(t)/C$ if $t \geqq \varepsilon$, $h(t) = 0$ if $t < \varepsilon$. Note that $C > 0$ for sufficiently small ε ($\varepsilon = 0$ is not an excluded value if $\sigma(0+) < \infty$), and that h is a probability density. It is now easy to compute η_s^2 as a compound Poisson distribution

$$\eta_s^2 = \sum_{n=0}^{\infty} e^{-Cs} \frac{(Cs)^n}{n!} H^{n*}$$

where H is the measure with density h; η_s^2 is absolutely continuous except for an atom at 0, whose value is e^{-Cs}.

If $\sigma(0+) < \infty$, we take $\varepsilon = 0$ and conclude that η_s ($= \eta_s^2$ in this case) is absolutely continuous except for an atom at 0.

If $\sigma(0+) = \infty$, since the convolution of an arbitrary distribution and an absolutely continuous one is absolutely continuous, we find that the mass of the not absolutely continuous part of η_s is at most e^{-Cs}. We now let $\varepsilon > 0$ tend to 0, C tends to ∞, and η_s thus is absolutely continuous. This implies the absolute continuity of E, and part 3) of Lemma 6 follows.

REFERENCES

[1] Chung, K. L.: *Markov Chains with Stationary Transition Probabilities.* Second Edition; Springer-Verlag, 1967.

[2] _____: "On the boundary theory for Markov Chains". *Acta Math.,* 110 19-77 (1963).

[3] _____ : "On the boundary theory for Markov chains". *Acta Math.,* 115 111-163 (1966).

[4] _____: "Markov processes with infinities". *Markov Processes and Potential Theory;* John Wiley and Sons, 1967.

[5] Doob, J. L.: "Topics in the theory of Markoff chains". *Trans. Amer. Math. Soc.,* 52, 37-64 (1942).

[6] _____ : "Markoff chains—denumerable case". *Trans. Amer. Math. Soc.,* 58, 455-473 (1945).

[7] Dynkin, E. B.: "General boundary conditions for denumerable Markov processes". *Teor. Veroyatnost. i Primenen,* 12, 222-257 (1967).

[8] Feller, W.: "On the integro-differential equations of purely discontinuous Markoff processes". *Trans. Amer. Math. Soc.,* 48, 488-575 (1940); Errata 58, 474 (1954).

[9] _____ : "On boundaries and lateral conditions for the Kolmogoroff differential equations". *Annals of Math.,* 65, 527-70 (1957).

[10] Kendall, D. G. and G. E. H. Reuter: "Some pathological Markov processes with a denumerable infinity of states and the associated semigroups of operators on ℓ". *Proc. Intern. Congr. Math. Amsterdamn* 1954, Vol. III, pp. 377-415.

[11] Kolmogorov, A. N.: "On the differentiability of the transition probabilities in homogeneous Markov processes with a denumerable number of states" (in Russian). Učenye Zapiski MGY, 148, Mat. 4, 53-59 (1951).

[12] Lévy, P.: "Systèmes markoviens et stationnaires. Cas dénombrable". *Ann. Sci. Ecole Norm.* Sup. (3) 68, 327-381 (1951).

[13] Neveu, J.: "Lattice methods and submarkovian processes". *Proc. Fourth Berkeley Symposium on Math. Stat. and Probability,* Vol. II, pp. 347-391, University of California Press, 1961.

[14] Pittenger, A.: "Boundary decomposition of Markov processes". Dissertation, Stanford University (1967).

[15] Reuter, G. E. H.: "Denumerable Markov processes (II)". *Journal London Math. Soc.,* 34, 81-91 (1959).

[16] Walsh, John B.: "The Martin boundary and completion of Markov chains". *Zeitschr. für Wahrscheinlichkeitstheorie,* (to appear).

[17] Williams, David: "On the construction problem for Markov chains". *Zeitschr. für Wahrscheinlichkeitstheorie,* 3, 227-246 (1964).